HARDCORE INVENTING

The IP³ Method:

**INVENT,
PROTECT,
PROMOTE,
AND PROFIT
FROM YOUR
INVENTIONS**

Robert Yonover, PhD and **Ellie Crowe**

Illustrations by **Micah Fry**

Foreword by **Louis Zamperini**

Skyhorse Publishing

Skyhorse Publishing books may be purchased in bulk at special discounts for sales promotion, corporate gifts, fund-raising, or educational purposes. Special editions can also be created to specifications. For details, contact the Special Sales Department, Skyhorse Publishing, 307 West 36th Street, 11th Floor, New York, NY 10018 or info@skyhorsepublishing.com.

Skyhorse® and Skyhorse Publishing® are registered trademarks of Skyhorse Publishing, Inc.®, a Delaware corporation.

Visit our website at www.skyhorsepublishing.com.

10 9 8 7 6 5 4 3 2

Library of Congress Cataloging-in-Publication Data is available on file.

Cover design by Rain Saukas
Cover photo credit Thinkstock

Print ISBN: 978-1-62914-426-9
Ebook ISBN: 978-1-62914-978-3

Printed in the United States of America

Contents

Foreword
by Louis Zamperini

Louis Zamperini, a 1936 Olympian and WWII U.S. Army Air Corps bombardier, survived forty-seven days at sea in an open raft and two years of torture as a prisoner of war. He was an inspirational speaker on life skills and forgiveness and coauthor, with David Rensin, of Devil at My Heels. *The major Hollywood film* Unbroken, *directed by Angelina Jolie and based on the bestselling book by Lauren Hillenbrand, depicts the events of his life. During his ordeals, "Lucky Louie" invented in order to survive.*

I FIRST CROSSED PATHS WITH DR. ROB YONOVER IN 1998. He'd been watching a television broadcast of the Winter Olympics, which were held in Nagano, Japan, and had seen me carrying the Olympic torch through a city in which I'd once spent miserable years as a prisoner of war. After listening to a CBS interview with me about my experiences of survival at sea, he located my telephone number and called me.

"I saw your interview and heard your story of survival—it was awesome," Rob said. "I'm an inventor from Hawaii. You know when you were on a life raft lost at sea and they flew over you? I've invented a long orange streamer that would have given you a tail so they could see you."

To me, that sounded pretty good! "Where were you fifty years ago?" I replied. Rob's inventions would have been a great addition to that life raft bobbing unseen for forty-seven days in the middle of the Pacific Ocean. I've got to say that the strongest emotion in human experience is going down in a plane, knowing you're going to die. A close second is when you're adrift in the open ocean and the rescue planes fly right over you and don't spot you!

Inventing is close to my heart. Even as a rebellious kid I was pretty inventive, though perhaps not in the best ways. My hook-and-toilet-paper invention to obtain coins from pay-phone slots worked well, and my "slightly used" tobacco made from unraveled cigarette butts and sold in Prince Albert tins was quite popular with unsuspecting pipe smokers! My milk bottle, coated white inside to create the appearance of milk, but actually containing beer, was a definite success with underage drinkers.

As a U.S. Army Air Corps bombardier during WWII, I needed a few inventing miracles when my plane ditched over the Pacific Ocean and I drifted, with two other survivors, 2,000 miles over the course of forty-seven days on a raft. When we eventually found a perfect tropical island, unfortunately occupied by the Japanese enemy, the other prisoners of war and I constantly relied on our inventive minds to survive torture and humiliation.

The RescueStreamer would have really helped me in 1943, as would Rob's DeSalinator for making drinking water. The life raft that my two air force buddies and I called home for forty-seven terrible days was woefully inadequate for survival. It lacked the basic necessities of a fishing kit and knife, the designers having strangely elected to give survivors at sea a pair of pliers instead! We improvised by using the pliers to fashion sawlike teeth on one corner of our chromed-brass mirror. Though sharp enough, this "knife" took ten minutes to saw through a shark's tough belly. Hungry men become quite ingenious. We used an empty flare cartridge as "bait." Once a shark had latched on to the cartridge, it instinctively wouldn't let go, and would become part of our food chain instead of vice versa. The shark's liver made a luscious, gooey meal.

One of the things that kept us going at sea was inventing fantasy meals. I'd invent a menu and describe the cooking process that I'd learned by observing my Italian mother at work in the kitchen. My starving companions would want every detail. If I omitted a detail, they'd pounce. "You forgot to grease the skillet!" or "What about the butter? Don't you need butter in gravy?"

After drifting 2,000 miles, our small life raft and its starving occupants eventually reached the shores of enemy-occupied Kwajalein Island. I weighed sixty-seven pounds. In the prisoner-of-war camp there was little food for the prisoners, many of whom looked like skeletons. We invented as many ways as we could to pilfer food. I admired the creativity of the Scots who worked at the enemy's warehouse. They specialized in smuggling sugar. They'd tie off their pant cuffs and fill the legs. Or they'd ask for bigger work boots so they could fill the boots with goodies.

Their most creative inventing took the form of excessive tea drinking. The Scots, to further the war effort, would drink tea all day and then take turns peeing on the enemy's rice so by the time it reached its destination it was spoiled. They took great delight in peeing on the cans of oysters addressed to the German chancellor!

I like the basic concept Rob expresses in *Hardcore Inventing*: Often what you need is already out there in the natural world, plain enough to see for anyone who is paying attention. Nature has always been an abundant source of inspiration to me.

Often the inventing process involves using existing things to create new ones. When I was a troubled youth, I learned to run fast by speeding away from the police. My brother, Pete, found a creative way to use that running skill—he introduced me to track, and this saved me from a life that was going nowhere fast. I became a runner and an Olympic athlete.

Today, I am a grateful survivor: I was forced to invent in order to survive. Rob's inventions for survival at sea were fifty years too late for me, but will help save the lives of others. Perhaps if survivors in life rafts are equipped with his inventions, they'll have more time to work on their fantasy dinner menus!

I hope this book helps all inventors succeed in finding and working on something they are passionate about and bringing the gift of their invention to the rest of the world.

Introduction
by Dr. Rob Yonover

GREETINGS, FELLOW INVENTORS AND CREATORS! A LOT has happened in the inventing world since the first edition of *Hardcore Inventing* debuted in 2009. First of all, I realized that "inventing" is an incredibly broad term. The principles in *Hardcore Inventing* apply to all aspects of creativity. Inventing, starting a new business, lifestyle changes, art, crafts, nutrition programs, exercise, and anything else that requires harnessing your mind to attack and solve problems is well served by the IP³ Method—Invent, Protect, Promote, and Profit—that forms the core of hardcore inventing.

My mantra remains "never quit your day job so you can quit your day job." Don't count on your creation paying the bills, especially right away. And even if things are going well, money brings in the sharks.

Recently, I personally had an experience of this when my inflatable paddleboard invention took off wildly in the Stand Up Paddleboard (SUP) market. Numerous companies started making inflatable SUPs and thus needed to sign up for a license and pay royalties to me, the patent holder, to obtain rights to use the patented technology.

As you might have guessed, large companies are not too happy with paying some hardcore inventor for the rights to a product they are selling with their names on it. Fortunately, I'd followed the roles in the book and protected my invention with a solid patent. To enforce it, I teamed up with a brilliant group of patent enforcement attorneys who have helped me vigorously defend my patent rights as an individual. In fact, the U.S. Patent Office has now acknowledged once again the uniqueness of my invention during a patent reexamination procedure, introduced by the companies that didn't want to see my patent rights exist. With minor modifications to the claims, my patent was upheld and even strengthened by successfully surviving the patent reexamination challenge!

While it's more important than ever to protect your invention, in the field of patents there is a major change that inventors need to pay close attention to. The recent major patent law change awards the inventor that is "first to file" with the spoils. This means that if you lose the race to the patent office, you have likely lost the ability to secure a patent. The old standard was "first to invent;" however, that is no longer true, and you are legally out of luck if you are not the "first to file" the patent application. My advice to inventors is now very simple: keep all of your work completely secret until you file (i.e., no public disclosures or presentations of your invention) and file your application quickly.

Unfortunately, this law change adversely affects the small inventor as the several-thousand-dollar typical cost of obtaining a patent will now be even more dominated by large companies that can afford to throw big money around. On the other hand, the inexpensive and sometimes equally effective methods of protection laid out in detail in the "Protect" section of *Hardcore Inventing* can also be extremely valuable in building an intellectual property portfolio around a new invention or creation (e.g., provisional patents, trademarks, trade secrets, domain names, copyrights, etc.).

There has also been a dramatic increase in the start-up investing communities and crowd funding sites that can provide you with both exposure and potential funding. Remember to pay close attention to the strings that come with exposure and funding, i.e., grants are free, loans have to be paid back, investors get a piece of your company, and crowd funding usually entails providing various perks to the investors.

Watch for grants from big corporations. They need inventors. Unilever has released a new set of challenges via its Open Innovation online platform. The company is asking inventors across the globe to help find technical solutions to three problems that it says will reduce the environmental impact of its more than 400 brands. One of these is for a non-artificial red color dye, hopefully to take the place of the ubiquitous Red 40.

The Bill & Melinda Gates Foundation has been steadily offering grants to sanitation initiatives, including some focused on creating better toilets for the developing world. Caltech won the Gates Foundation's Reinvent the Toilet Challenge (a call to create cheap, safe, and hygienic waterless toilets) with a solar-powered, self-cleaning toilet that converts urine and waste into hydrogen and fertilizer.

The Internet and social media explosion have provided many new outlets to promote your inventions; however, it has also astronomically increased the amount of competition of getting your message out. Ironically, trade shows, the thing I love to hate (see Chapter 13), remain the best way to get immersed in the specific sector or niche that your creation falls into. You can still get valuable face time with the players involved. You can always start small and tap local media and events to get some exposure for your invention.

Being an inventor remains one of the most rewarding things a person can do on any level, regardless of the success of your creation. You may not be granted a patent, but maybe you solve a problem that helps your business or makes your life a little better. The U.S. economy needs more people with great ideas, ones they are passionate about, to keep the U.S. job-creation machine humming. Check out the fields other inventors are working in and be inspired. The race is always on for the next product that will reign supreme. Exciting fields highlighted in the first edition of this book are constantly attracting new inventions as creative people build on what is already out there. Here are a few new inventions that may inspire you:

Digital Technology
MakerBot's new Replicator 2 can print out 3-D models with precision that's measured in micrometers. With the Replicator, you could take a 3-D model and bring it into the physical world for cents. Very useful for inventors who need prototypes; it may not be long before there's one of these in every home for making simple items like mugs and gears or more complex things like circuit boards and prosthetics.

Tech entrepreneurs predict big leaps in digital technology as the digital domain enters wearable technologies. Be inspired by Google's augmented reality glasses, watch and wristband pedometers, mood sweaters, fitness bracelets, anklets, and necklaces that track calorie burning.

Apps
Soon our homes, our cars, and even objects in the street will interact with our smartphones. On the positive side, this represents a huge market to

attack and solve problems in as an inventor or creator. On the negative side, these devices have, in my opinion, diminished the ability to discover new things personally. I still recommend going into nature or a quiet room with no electricity or devices to solve your next problem or come up with your next invention.

As the app market is booming, that invention might be another creative new app such as Morning, a customizable app that lays out your day's events, weather, and even traffic, or Happy Hour Pal, which finds the best place to relax after work. Travel Organizer will help build your itinerary and share it with your envious Facebook friends. Games get better and brighter and faster too. There's lots of room for imaginative inventors here. Cameras by themselves or on smartphones are a big field as they become smaller and better and offer bonus features, such as life-logging cameras that automatically snap for you.

Clean Energy and Green Inventions

Thousands of individuals and companies across the globe work hard developing alternative energy solutions. There is always room for new inventions focused on energy saving and clean energy.

Imagine a solar panel that is just a coating, as thin as a layer of paint, that takes light and converts it to electricity. From there, you can picture roof shingles with solar cells built inside. Maybe soon solar will be cheaper than burning coal. With the potential of 3-D printers becoming household objects within the next twenty years, we could very well see a future where we can print off our very own solar panels.

Inventors in the field of green technology offer innovations of the why-didn't-I-think-of-that variety. Examples are PepsiCo's plastic bottles made entirely of renewable and plant-sourced raw material. The Natural Energy Laboratory of Hawaii Authority is using frigid deep ocean water for air conditioning in buildings. Inventor Peter Brewin has worked long and hard on his recirculating shower to save water. And the (Almost) Waterless Washing Machine, invented by Stephen Burkinshaw, uses one cup of water, detergent, and 44 pounds of plastic chips per load to remove and absorb dirt and even stains, thus saving both water and power. Stephen Katasaros has invented Nokero, the world's first solar light bulb, for the 1.6 billion people worldwide without electricity. This is lifesaving when you consider that fumes from kerosene lights are equal to being exposed to forty cigarettes a day.

Military

New inventions are constantly being produced for the military. Why didn't anyone think of body armor specifically designed for women before? Garnering praise is the Smart Ball, a wireless data-gathering grenade to toss into danger. Invented by an MIT student, this baseball-sized orb holds cameras that send data to mobile devices. A low-budget interesting invention is the Meshworm, a tough Dr. Who-type of monster made of wire and paper that infiltrates small spaces. What about surprising the enemy with your own little kamikaze Switchblade Drone that goes to battle in a backpack?

Sports

The cardboard bike is made from reinforced and recycled sheets strengthened by a honeycomb and birds' nest design and will sell for around $30. Nike has produced light and flexible Flyknit Racers, composed of a single layer of knit thread. And people are still trying ways to fly, one of the newest being batlike suits with flapping wings.

The world is longing for new inventions, so don't wait too long to pitch yours. We are all born inventors and innovators, and I encourage you to keep embracing that part of your DNA. Remember, as always, to stay hardcore!

Aloha,
Dr. Rob Yonover

INVENT

Success Story—The RescueStreamer Technology

Money never starts an idea. It is always the idea that starts the money.
—*Owen Laughlin*

An invention is one of those super strokes, like discovering a platinum deposit, or a gas field, or writing a novel, through which an individual . . . can transform his life overnight, and light up the sky.
—*Tom Wolfe*

THE RESCUESTREAMER, AN EMERGENCY SIGNALING device used by all branches of the U.S. military and onboard all U.S. Navy submarines, and the Self-Deploying Infra-Red Streamer (SDIRS) now being placed on fighter-jet aircraft worldwide, were my first really successful inventions. I invented the RescueStreamer technology to solve a tricky problem I hope I'll never experience—a plane crash at sea.

Living on the edge was a reality for me, a big-wave surfer and scientist who worked on active volcanoes in the middle of the Pacific Ocean. The edge sharpened during an interisland flight on a small airplane that began sputtering like a Volkswagen Beetle. It didn't help my confidence that the plane was rented.

What if the pilot had to ditch in the ocean below? As usual in precarious situations, my mind started racing with contingency plans. I knew I could

get out of the cabin and swim, but the metal plane would sink quickly. Viewed from thousands of feet up in the air, the vast ocean is quite impressive. How would search and rescue parties see us? We'd be, to paraphrase marine biologist Roger Hanlon, yummy hunks of protein swimming in the ocean.

How could I somehow signal the search planes that would, hopefully, be looking for us? I brainstormed, but couldn't solve the problem. We finally landed successfully, despite the engine noises.

But the problem continued to rattle around in my brain until a few weeks later when I flew to Florida. As the commercial jet approached Miami International, I spotted a strange sight in the ocean below: small islands wrapped in bright-pink plastic—the artist Christo's latest project. That was it—I needed a little piece of that pink plastic and I could solve my visibility problem in an open ocean and on land. It took years to figure out how to put the pink plastic in a form that would be compact when not in use, yet extend to create a large target and stay that way until search parties could visually locate it.

Like all inventors, I've been a problem solver my whole life. My first invention, when I was about six years old, was a tiny battery-powered engine that spun a small shaft and performed automatically a bodily function boys especially are fond of (no, it's not that). I then taped a drinking straw onto the rotating shaft and produced the first mechanized electric nose picker! Inserting the rotating straw into your nose did the job quite effectively. However, if you stuck it in too far, it made you sneeze!

Ideas come and go in my maniacal brain, but that piece of pink plastic never left—that's how I knew it was a winner. A swathe of this bright plastic would be the ideal shape to see from the air; however, I couldn't figure out how to make it stay rigid and outstretched. I started looking for things in nature that remained extended. A few examples were striking—the human spine, centipedes, and palm trees. They were made up of segments, with supports at the beginning and end of each segment. I just needed to figure out how to put segments on a piece of plastic.

The next eureka moment came while I was teaching oceanography at the Hawaii Pacific University. I was in a laboratory handing out pipettes (small, semi-rigid plastic straws for measuring liquids) when I realized that the solution to the segmentation problem was literally in my hand.

Then it was off to my laboratory (read "patio at the back of my house") for research and development. I built the first RescueStreamer device there while my wife yelled at me to stop playing with plastic and get a real job! Those were words she would later gladly eat.

I built the first streamer out of white plastic, because I couldn't get bright pink as a free sample. As my wife and some friends sailed out on our small boat to watch the sunset, I unfurled the white streamer and it started to bunch up in the waves. I thought I'd again failed to solve the deployment problem. Then, suddenly, the currents stretched the streamer and the pipettes ("struts") caused it to straighten out like a spinning helix—the most beautiful thing I'd ever seen in the ocean!

The next step was to make it pink. I acquired more free samples: My method was to get free samples from large manufacturers with toll-free numbers, letting them know I was working on something big that could ultimately equal big sales for them. However, I could never make those pink versions work as well as the original white version.

In science, it is best to keep some things constant when creating new versions. I couldn't figure out the problem and thought it was the composition of the new pink material, as the chemistry of extruded plastics can be very complicated. Another principle I learned as a scientist is to keep good notes and always catalog samples, whether volcanic rocks from the ocean floor or new versions of a device. Finally, after racking my brain, I went back to the original white streamer that I'd fortunately kept in a safe place. It turned out the problem wasn't the plastic's composition; it was how I attached the struts. Initially, I was either too cheap with the glue or in too much of a hurry (or both) and didn't glue down the whole length of the strut against the plastic film. In the newer pink version, I'd taken my time and glued the whole length of the strut. That was the difference. By leaving air space between parts of the strut and the film, water was able to pass through the device without pulling it underwater. I'm glad I kept my original!

I like to bounce ideas off people and gather as many opinions (data) as possible, prior to making a decision. I showed the streamer to a navy captain who lived across the water from me. In the original design, the streamer is stowed in a pouch that unfurls to become a hat for the survivor to prevent sunburn while he is waiting to get rescued. The hat looked a little like a pirate's hat from a party on Fire Island. The captain liked my idea. However, his first words were "Lose the pink and lose the hat and you have something." I took his advice, and now all RescueStreamer emergency signaling devices are international orange, the color deemed most easily visible in the ocean by the U.S. Coast Guard. And they don't come with hats, though I still think it was a good idea.

PROTECTING MY INVENTION

The gut-check part of the invention game is getting a patent. Patents done right, and by that I mean by a patent attorney who can write "legal claims" that are defensible in court, are expensive, about $5,000. You can patent your invention on your own using various books on the market, but the "legal claims" are worth paying for.

In the case of the streamer, I tried to get a patent on my own the first time, but failed miserably. I tried again a few years later using a patent attorney and with legal claims in hand. The irony of patents is this: "the simpler the idea, the broader the patent." If you think about patenting a complex system like a computer, you can see that the next person to come along can just change a few electronic components and the device is different and distinct. In the case of the streamer patent, using struts as support bars across the length of the streamer film, regardless of the composition of the struts, provides for a very broad patent because it is such a simple design.

In choosing a patent attorney, I chose a longstanding firm that was right across from the patent office. Proximity proved invaluable, as it is common for the patent attorney and the examiner to have face-to-face meetings as they debate how strong the patent will be.

PROMOTING MY INVENTION

Now I ran into that point in every inventor's history that I call a "critical crossroads." This is when you run into an obstacle—and I guarantee you will run into them. You have a few paths to choose from, with the easiest being quitting. I think inventors who take on a relentless persevering attitude are the ones who succeed. You will encounter people who laugh at you, make fun of you, blow you off, hang up on you—and these are just your family members! You have to drag yourself off the ground and get up and keep charging ahead.

To keep sane, I found the best approach to keeping the financial pressure off was to keep my day job. I had many day jobs in my quest to become a profitable inventor: teacher, housepainter, environmental scientist.

During my journey, I learned a critical lesson from my brother, who was involved in advertising and marketing: Every day editors of magazines and newspapers are sitting at their desks wondering what to write about. It takes a special type of person to pitch a story to 100 people and get rejected 99 times. In the end, you have to believe that there is one editor out there willing to write about your invention.

Before e-mailing, there was faxing. Late at night, when phone rates were cheaper, I'd send the same fax to 100 people. Finally, I got a sports editor at the *Miami Herald* to write about the RescueSreamer. Once the *Miami Herald* article came out, I mass-faxed copies of that article to the 99 other people and alerted them to how the big guys were writing about the streamer. I repeat the process to this day and jokingly refer to myself as a "media whore." The RescueStreamer has now been featured on CNN and the Discovery Channel, and in numerous magazines, including *Outside* and *Playboy*.

During my "don't quit your day job" period, I continued relentlessly self-promoting to the U.S. military, writing polite but compelling letters to the highest-ranking admirals and generals I could find. A few actually answered, leading to a U.S. Navy trial of my streamer. It passed with flying colors and I used that bit of news to again try to whip editors into a frenzy.

As a result of the publicity, a venture capital group learned about the streamer and invited me to submit a business plan to a panel of experts that included a marketing executive and a patent attorney. I was revved up as usual for any public appearance, particularly one with major players in attendance. The panel came at me from left and right, and I was ready to defend my baby. After I aggressively defended the RescueStreamer, the session ended with applause from everyone. An attractive woman approached me and identified herself as a reporter for a major weekly business publication. The result was an article alluding to massive profits in the streamer's future and lives to be saved from the newly approved military survival technology! Of course, I made copies and continued to fax them out to more publications.

PROFIT: WHAT ALL INVENTIONS DO IN A PERFECT WORLD

A few days later, a couple of local "angel" investors called me and I had what all inventors dream about—a bidding war! One group outbid the other and I had a better feeling about working with them. I essentially invented a licensing agreement that linked us together as closely as true partners without the legal risk of being true partners. (Profit sharing licensing agreement.)

The rest is history. The RescueStreamer is now approved and used by militaries all over the world, and every U.S. Navy submarine has a

RescueStreamer for each person. The new Self-Deploying Streamer (Automatic Day and Night Signaling) is approved and about to be placed under the butts of military fighter pilots all over the world! What's even better is that the RescueStreamer technology has already saved four lives and counting (two skin divers and two military personnel in battle)! That was the intention, to contribute something of substance to this world and to help humanity in the ways of my ultimate mentor—Buckminster Fuller! Securing a licensee was only the first step toward profiting from my invention. To ensure good sales, I still had to help the licensee "sell" the streamer. I continued to leverage any and all third-party endorsements I could get, in order to secure distributors, dealers, and customers.

I believe people are followers in general and if they see the Joneses buying something, they want one too. With numerous third-party endorsements, including some from highly respected parties like the U.S. Navy and the U.S. Air Force, expanding my customer base became increasingly easy. Another key to the success of the streamer was trade shows, because they put a little guy like me face-to-face with the power brokers. Coming from my humble abode in the middle of the Pacific Ocean, I follow a couple of rules prior to appearing in public at trade shows: Get a short haircut prior to departure, and overdress for success—when everyone is dressed in suits, we are on quasi-equal footing. Equipped with my thirty-second elevator pitch and an attitude of a polite, persevering pain in the ass, I attack targets relentlessly until I make a breakthrough. Also, I am not shy about playing one magazine against another or one customer against another—in the sense of getting the non-committed party to commit based on the actions of the other.

The bottom line is: Learn about your customers. Go where they live, work, and shop. Talk to them and to store owners. Watch them buying similar products. The customer is always right. I learn their requirements and rework my ideas. If the customer wants a yellow streamer instead of an orange one, as did the U.S. government, I give them that; it's all about relationships. And profiting from your inventions is a series of relationships, including those with your licensee, the patent attorney, the magazine editors, distributors, and finally the end customer. Never burn a bridge to any of them.

Identifying and Solving the Problem

Discovery is almost never a single idea. Always look for new, related problems after solving the initial one.

—George Polya, mathematician

Don't assume you have to be an expert or an insider to invent and innovate.
—Maurice Kanbar, inventor and author

INVENTIONS SOLVE PROBLEMS. MOST INVENTORS PROBLEM-solve their whole lives, constantly looking for ways to construct things more efficiently. To come up with ideas, I keep a close watch on the world around me, talk to people in fields other than mine, listen to problems, and think about how I can improve things.

K.I.S.S. (KEEP IT SIMPLE, STUPID)

People looking at my inventions often say, "Sorry to insult you, but that looks so simple." To me it's not an insult—it's a compliment. Intricate things have more components that might fail: Batteries go flat, electronics malfunction, chemicals change or dissipate. I approach inventing in the simplest, most primitive fashion.

USE YOUR INNER APE

Our ancestors roamed the earth as apelike creatures, surviving through primitive instinct. My basic philosophy is to return to those natural inclinations and simplify whenever I can.

Cavemen were the first inventors and they used whatever they had around them for materials. To survive, they had to be creative and they were. Europe's oldest natural mummy, nicknamed Ötzi the Iceman, lived around 3300 BC. Ötzi's well-preserved remains were discovered in 1991, buried in a glacier in Austria's snowy Alps. His surprisingly sophisticated clothing and equipment show the creativity of primitive man. He wore a cloak of woven grass and a leather belt, leggings, a loincloth, and wide, waterproof leather snowshoes. Self-sufficient Ötzi carried a fire-starting kit with flint and pyrite, medicinal mushroom fungus with antibacterial properties, an axe with a head of smelted copper, and arrows of flint with sharp, scalloped blades.

When I first attack a problem, I like to pretend I'm a caveman (or waterman, in my particular watery domain) and force myself to come up with a solution quickly, using only the materials at hand. The first order of business for a true primitive creature of the earth is to carefully observe his surroundings for ideas and inspirations.

LOOK TO NATURE FOR SOLUTIONS

You've probably looked around and been amazed at how the plant and animal world evolved to conquer potentially life-threatening or species-threatening obstacles. Living on an island in the middle of the Pacific Ocean has made this clear to me. Nature has answers that inspired some of the most astonishing breakthroughs in medicine, materials, and artificial intelligence. Always look to nature to see how problems are solved naturally and how things are constructed to withstand natural forces and threats. Inventions as diverse as aircraft, Velcro, and robots are examples of biomimicry—taking designs from nature. Flight is a perfect example.

From Birds and Trout to Jumbo Jets

For centuries, inventors who yearned to fly looked to nature and studied birds. Some early aviators actually copied birds' feathered wings, but failed to understand the nature of flight from an engineering point of view.

Leonardo da Vinci said, "There shall be wings! If the accomplishment be not for me, 'tis for some other." Unfortunately, his design showing a man in a huge wooden frame trying to flap his "wings" was doomed to fail because the amount of muscle required for propulsion was too great. Ultimately, it was an English inventor, Sir George Cayley, who realized the secret of flight wasn't to be learned from birds flapping their wings but by watching birds glide with their wings fixed.

Cayley, often called the "Father of Modern Aerodynamics," identified the three forces acting on the weight of any flying object as lift, drag, and thrust. Again borrowing from nature, Cayley noted that a heavy trout swimming through water provided the ideal minimum-resistance body shape for an airplane. By 1804, he had invented and was flying model gliders.

In 1903, the Wright brothers made the first controlled, manned, powered, heavier-than-air flight. These inventors noted that when birds fly they twist the back edge of one wing upward and the back edge of the other wing downward to change direction. They incorporated nature's manufacturing genius into the design of their flying machines.

From Burrs to Velcro

Many a successful invention comes from an amazingly simple premise. Ever hiked through a field and found dozens of prickly burrs attached to your jeans? Ever tried to get those suckers off? The inventor of Velcro, Georges de Mestral, noticed how tightly burrs stuck to the fur of his dog. Running to his microscope, he examined the burrs and noted tiny hooks. It occurred to him that this was a good way to join two fabrics together.

Mestral's idea met with resistance and even laughter. He persevered. Together with a weaver from a textile plant in France, he perfected his hook-and-loop fastener and formed Velcro Industries to manufacture his invention. Soon he was selling over sixty million yards of Velcro per year. Velcro is now a multimillion-dollar industry.

There is still room for inventing here: Velcro makes a noise when it's ripped apart. Modern military uniforms have lots of Velcro as fasteners, and noise is bad if you're a sniper! Scientists today are trying to invent silent Velcro and Vermont architect Leonard Duffy seems to have succeeded.

The Slidingly Engaging Fastener

After experimenting for eight years with foam, cardboard, and wood, Leonard Duffy came up with his new product, the "Slidingly Engaging Fastener."

Duffy earned an award at the 2007 PopSci Invention Competition, gaining needed publicity for his new version of Velcro. He also won a NASA-sponsored invention contest called "Create the Future." Nearly 150 product designers contacted him after the clearinghouse Material ConneXion added his product to its library last year.

Duffy is presently involved in licensing different applications of his invention. He has received e-mails from all over the world from people using one of these applications, a removable plaster cast called the Unitary Wrap. Check out www.lynxfasteners.com for information about Duffy's products.

From Geckos' Feet and Frogs' Toes to Nano Adhesive

Nature has engineered other brilliant ways of making things stick together: Tree frogs have tiny suction cups on their toes, and for traction snails secrete special mucus. The gecko, whose feet are not sticky or gluey, has the ability to adhere to surfaces as smooth as glass. In 2007, a team at Northwestern University unveiled a gecko-inspired adhesive combining the microscopic marvel of sticky gecko foot pads with the ability to adhere to wet surfaces unique to mussels, making a glue that, moist or dry, always sticks. This product will be used for surgical bandages and patches to cover damaged tissue.

From Fish and Lizards to Robots and UAVs (Unmanned Aerial Vehicles)

Creatures from fish to lizards have inspired a class of robots that can go where people can't, or prefer not to, go. Many of these autonomous vehicles are being developed by Boston-based IS Robotics with funding from the Office of Naval Research and the Defense Advanced Research Projects Association.

Robots called DARTs (Device for Acceleration and Rapid Turning) are based on the morphology of freshwater pike. With segmented bodies like the fish, DARTs use a similar undulating swimming motion that allows them to accelerate and turn rapidly. IS Robotics has developed a three-foot-long DART prototype in collaboration with the MIT Department of Ocean Engineering. This can navigate hydrothermal vents deep under the sea or conduct near-shore military surveillance.

Ever watch a crab scramble over obstacles? Ariel, a six-legged robot whose design is based on a crab, is used to remove mines and obstacles underwater and in the surf zone. Ariel can scramble over obstacles and crevices upside down and right side up.

Professor Mark Cutkosky, along with biomimetic designer Sangbae Kim and a team from Stanford University, has invented a cool robotic gecko named Stickybot with geckolike feet. The Pentagon has taken an interest in Stickybot—this robot could act as a spy, a James Bond–type gecko. Stickybot is another example of science mimicking geckos, whose widely spread toe pads, lined with millions of hairlike structures, can climb in any direction on virtually any surface and can move between horizontal and vertical surfaces with ease.

BigDog, the alpha male robot of Boston Robotics, is a robot husky the size of a humungous dog or small mule. BigDog's legs are articulated like an animal's legs and have shock-absorbing elements that recycle energy from one step to the next. BigDog has an onboard computer that controls movement and sensors, and a control system that keeps him balanced, steers him, and navigates as conditions vary.

BigDog can run at 3.3 miles per hour, climb a 35-degree slope, and haul 340-pound loads over harsh terrain as it carries equipment for the military. You wouldn't want to meet BigDog on a dark night in the woods, but you can see him there at www.bostondynamics.com/content/sec .php?section=BigDog. And he'd make an extraordinarily effective watch-dog! Good dog!

Boston Robotics is busy developing a range of mobile robots that include a flapping UAV the size of a dragon and critters that climb straight up vertical walls. Marc Raibert, the team leader, says they have projects so strange that "you wouldn't believe them, even if we could tell you about them." A James Bond–style research project is actually on the drawing board at Wright-Patterson Air Force Base in Dayton. Here, military engineers are developing Micro Aerial Vehicles (MAVs), disguised as insects like bumble-bees, that could one day spy on enemies and conduct dangerous missions without risking lives. The project is starting with the development of a bird-sized robot as soon as 2015, followed by insect-sized models by 2030. Enabling a robotic insect to carry the weight of cameras and microphones is a major difficulty. And real insects are able to bounce off walls and keep flying—these robotic insects should ideally be able to do this too. The MAVs could also carry chemicals and explosives into enemy buildings.

SUMMARY

- Look for a problem
- Listen to other people's problems
- Look for a solution
- Keep it simple
- Look to nature for solutions

How Nature Inspired My Inventions

Human subtlety will never devise an invention more beautiful, more simple, or more direct than does Nature because in her inventions nothing is lacking, and nothing is superfluous.

—Leonardo da Vinci

THE FOLLOWING ARE EXAMPLES OF HOW NATURE provided inspiration in the formulation of some of my better-known inventions:

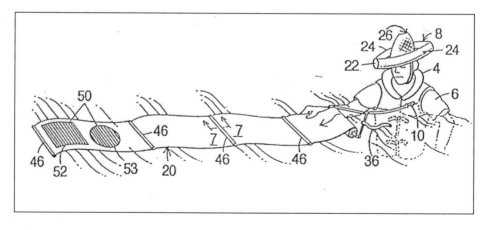

RescueStreamer

S.D.I.R.S.

RESCUESTREAMER TECHNOLOGY

When inventing the RescueStreamer technology, I was challenged by the problem of making a person or object in an expanse of water or land visible to rescue parties. I looked to nature by evaluating aerial pictures of water and land and by looking down to earth from an airplane. Although there are many colors and strange shapes, the existence of straight lines is very rare. The only true naturally occurring straight lines I saw were at the edges of cliffs.

Neuroscientists say the brain best responds to straight lines. In a series of tests, humans were subjected to a variety of specialized colors and shapes to see which stimuli most quickly triggered the brain. It was only when the photographic slide slipped off the viewer, exposing the straight edge of the photo, that the brain went crazy and fired off immediate responses! The scientists concluded it was likely something in our genetic makeup providing a warning in potentially dangerous situations.

On a more mundane level, if you drop a two-inch piece of sewing thread and a two-inch toothpick on your carpet, you immediately see the straight line of the toothpick before the irregularly shaped thread, especially if the toothpick happens to be a bright color. Thus the RescueStreamer concept was born; however, it still had to be constructed to work properly. I needed a straight, floating, orange line in the ocean.

Once again, I looked to nature for examples of elongated objects that remain relatively straight despite outside forces. The first two that came to mind were the human vertebrae and a palm tree. The common denominator in both objects is their segmented structure. I separated the streamer into segments by affixing air-filled struts at a constant spacing, which also added to the flotation capability.

The segments enabled the streamer to right itself and kept the orange film linear for maximum surface area to be quickly and easily identified by rescue parties.

I still had to make the RescueStreamer visible at night. Nature helped again—this time 10,000 feet down on the Galapagos submarine volcanic ridge. As a scientist working on Hawaii's volcanoes, I'd gone to the ocean floor on the ALVIN submersible, one of my wildest dreams come true. As the metal titanium sphere descended, the water changed from blue to green to gray to black, and glow-in-the-dark creatures rained down around us.

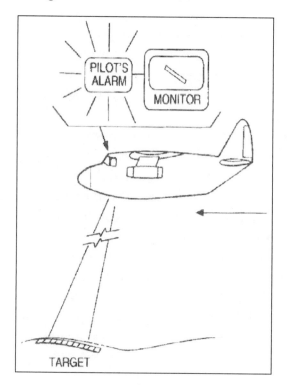

vSAR

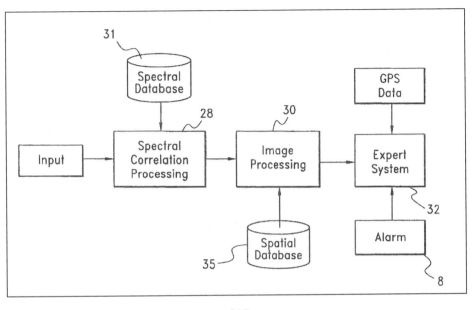

vSAR

Upon touchdown, I peered through the three-inch windows out at the ocean bottom, a moonscape with a fault zone showing a deep, dark crack wider than the sub. (The pilot insisted on staying a long way away from that!) Tiny plants, like small palm trees with flowers on top, swayed to and fro. Their thin, segmented stems bent with the strong currents but remained intact, validating my choice of palm tree–like struts.

Being the low man on the totem pole, I worked the night shift dredging for volcanic rocks in the pitch darkness. Strange critters roamed these depths. One animal looked like a monkey's skull with a clear, eel-like tail, its inner organs glowing in a variety of colors. I'd found the perfect solution for a visible-at-night RescueStreamer—glow-in-the-dark bioluminescent struts!

VSAR TECHNOLOGY

From a primitive perspective, humans are clearly not the best-endowed animals in terms of eyesight. I repeatedly noticed during my rough-water fishing expeditions how well birds see things from above. Not only can they see from an altitude similar to that of human search parties, but they also have much better attention spans when looking at an endless view of blue water.

I'd heard about a military project called SeaHawk. Pigeons were trained to peck on a lever if they saw the color orange, resulting in birdseed for them and a warning to the pilots that an orange object was bobbing in the ocean below (orange being the typical color of life jackets and life rafts). The pigeons were strapped inside a bubble mounted on the bottom of a helicopter. The program worked well despite the mess the pigeons made when the birdseed came out the other end. I decided that if I were lost at sea, ideally with a long orange RescueStreamer trailing behind me, I'd want something with birdlike vision looking for me.

My subsequent invention was vSAR (Video Search and Rescue) technology. The vSAR technology was the result of brainstorming with a couple of other egghead PhD types, one a high-technology camera expert who had worked with NASA and Carl Sagan and the other a computer software

Pocket Float

expert who had trained with a mad Russian physicist. Using multispectral cameras combined with automatic-target-recognition computer software, we invented the vSAR technology. VSAR uses cameras to locate specific colors (spectral signals) and shapes (spatial signals) to automatically locate targets of interest by setting off an alarm.

Thinking about the trailblazing pigeons, I adopted the pet name "the electronic pigeon" for the vSAR technology in their honor. The vSAR technology received a military grant for development and is now patented. I am presently working on the next generation vSAR, where the technology will be integrated with unmanned aerial vehicles and robot boats with the ability to "see."

If I were lost at sea, I'd want a RescueStreamer device trailing behind me and "birds" looking for me in the form of airplane-mounted cameras with automatic-target-recognition computers. We already see the evolution of air operations moving toward unmanned aerial vehicles (UAVs), which are essentially artificial birds flying around with computers for brains— another example of humans working extremely hard and making major advancements, only to arrive at the level that nature already inhabits!

POCKETFLOAT TECHNOLOGY

In continuing my saga of rescuing the lost-at-sea, it occurred to me early on in my ocean-based adventures that flotation is the first step to survival. If you can't breathe underwater, you have to be able to float on the surface to survive. Humans don't naturally float. Also, they are reluctant to wear bulky life jackets while they are out on the ocean, as humans are macho with an it's-not-going-to-happen-to-me attitude.

One creature that fascinated me was the blowfish, or puffer fish, which escapes danger by inflating its body and floating to the surface, making itself larger and thus more difficult to eat (its sharp outer spikes also help). I combined the blowfish expanding and rising to the surface (built-in, compact, inflatable ballast) with the phenomenon of air bubbles at the surface of the water. After a wave breaks, numerous tiny spherical air bubbles form at the surface of the ocean. These bubbles are incredibly resilient to the wind and water currents. Their spherical shape enables them to skip and slide along the surface, keeping their shape until they ultimately pop, releasing the captured air.

With blowfish and air bubbles floating around my head, I recalled the survival saga of a navy seaman who was lost overboard. Instead of tying a knot in his pants and using them to capture air for flotation, as taught by the navy, this inventive guy remembered that he had a condom in his pocket. He decided it would be better to inflate a condom for flotation once than to have to take off his pants and inflate them repeatedly. (Most

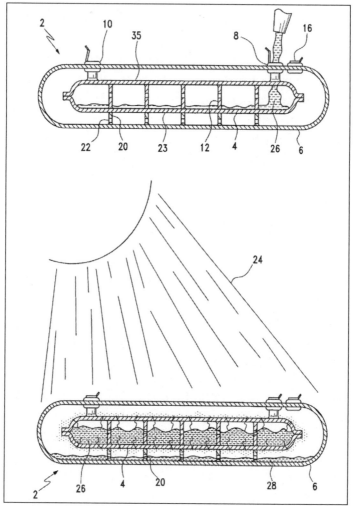

Desalinator

men don't want to be pants-less with sharks roaming below!) He held onto the inflated condom and it kept him afloat until he was rescued.

I thought that was brilliant and immediately designed a system with two condoms (inflatable ballasts) and a strap to place them under the person's arms for hands-free flotation. I started with condoms since they proved successful already and few balloon structures are as thoroughly tested before and after purchase!

The concept started out as a party joke; however, the more I thought about it the more I started to believe it would make for an excellent survival tool. Of course I did not use condoms, but rather meteorological balloons for the second prototype. (After Wal-Mart punching-ball balloons, which were the first.) The technology has now been funded by a military grant; a U.S. patent is in place, and manufacturing is underway using high-performance plastics instead of latex balloons, and lights and radio signaling are additional capabilities added to the technology.

DESALINATOR TECHNOLOGY

One of the most limiting problems humans have in any extended stay in saltwater is the lack of freshwater for hydration. Water is more important than food when it comes to survival. Given that our survivor at sea, using my inventions to date, now has the capability to float, signal, and be detected by search parties, this will all be for naught if the survivor doesn't have water to live. But water is one of the bulkiest and heaviest things to carry.

During many boat excursions, I noticed that evaporation occurred extremely quickly and completely on aluminum surfaces or other surfaces that get hot; i.e., drops of water on such surfaces get quickly converted to gas, leaving only salt residue behind. Solar stills can separate freshwater out of saltwater by evaporation and then condensation; however, they are usually bulky contraptions that proceed at a slow pace. Drops of water on a window screen remain larger than the holes in the screen because of the surface tension inherent in a sphere of water. I wondered whether there was a screening with openings so small that salt crystals couldn't pass through but water vapor could.

Out in a boat on a stormy day, I happened to have taken my Gore-Tex jacket along and was amazed at how well it worked as a filter, letting air in and out to let skin breathe without getting wet (water droplets could not fit through the filter). Gore-Tex is essentially a web material consisting of tiny openings that are too small to let liquid water through, yet large enough for water vapor (gas) to pass through.

The proverbial lightbulb went off in my head, and I wondered if I could use Gore-Tex-type material as a filter to trap salt solids and separate them from saltwater. Once again through a military-funded grant, I came up with a double-plastic-bag design, with the inner one being black-colored and containing a patch of Gore-Tex and the outer one being clear. The idea was to load saltwater into the inner black bag (evaporation bag), which is then heated up by the sun, causing the water vapor of the saltwater to separate out as freshwater vapor that can pass through the Gore-Tex, leaving the salt crystals behind. The outer bag (collection bag) traps the water vapor and then condensation takes over and the vapor is converted back to liquid for drinking.

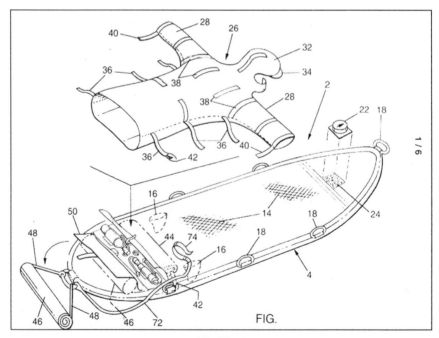

FIG.

Life/Float

The DeSalinator technology is not only useful for survival applications, but can be used to generate drinking water in larger quantity for emergency situations such as a water source gone bad during hurricanes or in third world countries without safe drinking water. The larger DeSalinator version resembles a Slip-n-Slide, the children's slide made from a long, wet, plastic strip. Once again the simple, natural processes of evaporation and condensation led to solving a longstanding problem for humans, even though nature had solved it eons ago: Many animals search for naturally occurring vaporization/condensation situations for freshwater ingestion.

LIFE/FLOAT TECHNOLOGY

The final technology in my transition from human to waterman consists of providing the human with the means to exist for extended periods of time on the water (until the rescue parties find you or you are ready to return to civilization, whichever applies).

Insects and other animals that crossed great oceans and propagated their species needed a floating object to survive, and humans on the surface of the ocean need flotation and mobility. In the survival world, a one-man life raft looks like a hot dog bun, with the survivor lying in the middle and drifting helplessly until help arrives or he becomes the meat of the hot dog for a shark. I wouldn't want that fate. I'd want to have the ability to move toward help, move toward a land mass, move toward a fish to catch, etc.

Drawing from my surfing background, I noted that humans are equipped to move through water by paddling their arms while lying on top of a floating object. In constructing a durable and rigid flotation device, I reflected that connecting tendons and muscles reinforce the most resilient human body parts and that there must be an artificial product that uses this concept.

Sure enough, in the 1940s, life raft manufacturers had come up with a concept: the outer layers of a raft floor connected by thousands of interior strings that could be inflated to very high pressures to make a rigid structure capable of supporting the weight of a standing human. This material was referred to as "drop-stitch" construction, referring to the thousands of strings connecting the two pieces of material together (i.e., they were not stitched together tightly, but thread was left hanging to create a void where high pressures of air could be filled).

Again with a military grant, I combined the design of a surfboard with the drop-stitch material to create a rigid inflatable surfboard that could support the weight of a human, even standing up, yet was compact enough to fit into a backpack when not in use. I built in a thermal protective cover (wetsuit-type structure) and a RescueStreamer to provide mobility, thermal and solar protection, and emergency signaling. The patented LIFE/FLOAT can be used as a life raft or it can be dropped to flood victims to provide a means to escape floodwater or transport injured humans or personal property.

In transitioning from human to waterman, I have assessed the requirements of existing in the ocean world and launched the aforementioned technologies, all based on naturally occurring systems. With a pack the size of a large backpack, a human could be dropped into the ocean anywhere in the world and have a LIFE/FLOAT and a PocketFloat for flotation and mobility, a RescueStreamer for emergency signaling, and a DeSalinator to produce drinking water. Of course, all LIFE/FLOAT packs are equipped with a fishing kit for your dining needs. On the other end, the rescue parties coming to look for you, whether they are humans in a plane or unmanned planes, will be equipped with vSAR to make sure they find you before the sharks do!

Scour Your Brain

People all over the world are coming up with inventive new ideas, and they're using the Internet and new communication tools (whose prices are falling every day) to share these ideas more quickly and more richly than ever before.
> —*Evan I. Schwartz, author*

Invention is 99 percent perspiration and 1 percent inspiration.
> —*Thomas Edison*

WAITING FOR THE INSPIRATION FOR A GREAT PRODUCT? Try looking at other inventions.

Ralph Waldo Emerson said an inventor knows how to borrow, and successful innovations rarely come from the person who first conjured up the basic idea. Today's inventors combine different ideas from past innovations. Success stories contain a wealth of information for new inventors.

Success Story—IllumiNITE

In 1989, while students at the Sloan Management School at MIT, Adam Rizika, Bob Rizika, and Scott Brazina dreamed of starting a business together and introducing the "next big product." But they had no idea what that product would be.

In 1993, they started Reflective Technologies Inc. and commenced searching for the right product. The partners planned to visit the technology licensing departments of major universities to see what technology was already available. However, they found that Adam's employer, Airco Coatings, had rejected a lot of outside technology. Among the rejects was a new type of reflective technology.

The partners checked the potential market for the product and liked what they found. They bought the inventor out and spent the next year developing the formula and technology behind IllumiNITE. Next they applied for a patent. They generated initial market contacts, taking material samples to companies like Nike and asking how they felt IllumiNITE could be incorporated into their product line.

To choose their marketing strategy, the partners decided to model their business after Gore-Tex, the very successful company that makes waterproof fabric. They also kept a close relationship with the Sloan Management School.

Today, Reflective Technologies Inc. is used by the Sloan Management School as an example of the classic case study involving:

- Finding technology that's hard to copy, easy to implement, and able to generate high profit margins
- Obtaining existing customers who are dissatisfied with current products
- Discovering competitors with outdated technology
- Producing a product that is easily understood by customers (IllumiNITE concentrates the light it receives and reflects it back so the wearer can be seen at a distance at night, but the product doesn't look that reflective up close. The company compensated for this with its packaging by using strong visuals of a runner wearing the reflective fabric.)
- Finding a product that offers a good price/earnings ratio

From their small beginnings, the partners have created a multimillion-dollar company with such major clients as Land's End, Adidas, Polaris, Eddie Bauer, Honda, L.L. Bean, and the U.S. military.

Webkinz by Ganz

Imitation is the sincerest form of flattery for inventors in the virtual world too. In 2005, Ganz created Webkinz, cute stuffed animals with a unique feature: Each toy creature comes with a code that unlocks its virtual double in a Webkinz virtual world at www.webkinz.com. This was obviously a great idea and the value of the Webkinz brand reached $2 billion in three years, inspiring a growing number of imitators who are unleashing their own toys, each with a unique feature giving access to virtual worlds.

Disney has a Web site based on the Pixar movie *Cars*. Here, Hot Wheels cars race special tracks and players can buy upgrades with a "race card" that gives online access.

And Barbie is now in her own virtual world, accompanied by players wearing sparkly tiaras. Avatars gain access to the world through a Barbie MP3 player plugged into a PC. Mere commoners can cruise this world too, but they pay a fee and don't get to wear a sparkly tiara.

BRAINSTORM

As an inventor, you will have your own ways that work best for creative thinking. Getting the initial idea is like an engine getting a spark to start. It sets off a series of reactions, one of which is brainstorming for the best way to implement the idea.

Brainstorming with a trusted group of creative people is a great way to come up with ideas or solutions to a problem. Thomas Edison and his crew of fellow inventors amassed 1,093 patents as they worked with motors, chemicals, and machines, and invented the lightbulb, the movie camera, the phonograph, and more than a thousand other innovations. These colleagues spent years freely corresponding and brainstorming with each other, exchanging ideas that later became the foundation of modern physics. They avoided bickering by applying a method of discussion called *koinonia,* or spirit of fellowship, established by Socrates.

KOINONIA OR SPIRIT OF FELLOWSHIP

1. Establish dialogue.
2. Exchange ideas without trying to change the other person's mind.
3. Don't argue.
4. Don't interrupt.
5. Listen carefully. Focus on who is speaking.
6. Clarify your thinking. Suspend all untested assumptions and look at everything with an unbiased view.
7. Be honest. Say what you think even if your thoughts are controversial.

USE TECHNIQUES FOR CREATIVE THINKING

The S.C.A.M.P.E.R. technique for creative thinking, created by Bob Eberle and discussed by Michael Michalko in his book *Thinkertoys, A Handbook of Creative Thinking Techniques*, will assist you in thinking of changes you can make to an existing product to create a new one. (See more at www.creativethinking.net.)

S.C.A.M.P.E.R

Substitute: What can you use instead of the ingredients, materials, or methods now used? Examples: Diet Coke, helium balloons, paper napkins.

Combine: How about blending or combining purposes? Examples: cell phones with cameras; printers with scanners and fax machines; clocks with radios, telephones, and tape recorders.

Adapt: What else is like this? What other idea does this suggest? Examples: soap containers shaped like shells, lollipops shaped like pacifiers.

Minify: Can it be smaller or lighter? Examples: smaller iPods, tiny cell phones, smaller laptops.

Magnify: Can it be bigger or stronger? Example: fifty-two-inch LCD HD televisions.

Put to other uses: Is there a new way to use it? Zippers were first used as fasteners on boots. Other places to use it? Examples: snowboards, mountain bikes.

Eliminate: What can you remove? Examples: decaf coffee, bread without trans fats.

Rearrange: Can you use another layout? Change pace? Can you turn parts backward, inside out, or upside down? Examples: reversible clothing, Gore-Tex used in the DeSalinator.

Reverse your perspective. Ask: "What is the opposite of this?" Find a new way of looking at things. Henry Ford's invention of the assembly line brought the work to the people. Ford said, "I discovered nothing new. I simply figured out how to assemble a car faster and better than anyone else."

VT2: VIRTUAL THINK TANK

One of the ultimate concepts in creative, scientific, business, or social endeavors is the "think tank." In a hardcore, bare-bones world, it is a good strategy to create a "Virtual Think Tank" (VT2) with one or more experts or confidantes. At the most basic level, you can be a one-person Virtual Think Tank and try to troubleshoot or play devil's advocate to analyze your ideas, actions, or plans. Try to enlist family or friends to take part in your VT2, without them knowing they are part of it. I like to bounce things off everyone and create instant dynamic VT2s, some of which may last, depending on what you get out of them.

Trust is important in VT2s, so be careful to only broach the sensitive issues with the people you can really trust, usually family and some close friends.

I bounce things off my father, brother, and wife all the time. I know their predispositions to things and that helps in bringing up the subject matter for a specific VT2. For instance, if one family member is very hard-nosed in business, I will save a topic for him that will lead to lively, yet useful, banter. Don't be afraid to get into heated arguments and debates, because that will force you to take a hard look at your position.

The act of debating subjects leads to a no-nonsense attitude and forces you to think on your feet, reexamining your opinions or actions from another person's perspective. Plus, using your family and friends as VT2s saves you a hefty consultation fee.

WHAT'S GOING ON OUT THERE? TODAY'S HOT INVENTIONS

While scouring your brain for ideas as to what to invent and what is needed out there, it helps to keep an eye on what today's inventors are inventing. According to Jeffrey Dollinger, chief development officer of the National Inventors Hall of Fame Foundation Inc., there is presently a trend among inventors focusing on medical, safety-oriented, and environmentally friendly innovations, and computer technology.

Medical Inventions

SUCCESS STORY

SimpleShot

Kim W. Bertron and co-inventors Brian J. Boothe and John Wiley Horton invented SimpleShot, a medical device that simplifies the process of mixing a powder-form drug with a mixing solution in a single syringe. In an emergency situation this device provides faster, easier administration of reconstituted drugs.

Bertron's invention was the result of a medical emergency. She and her ten-year-old daughter, who has type 1 diabetes, were on a beach vacation when her daughter suffered a severe hypoglycemic episode, including a violent seizure, and became unconscious. Bertron needed to administer a lifesaving dose of glucagon. As she was frantically trying to mix the drug in powder form with the diluting solution, she broke the needle on the vial. Fortunately, Bertron was able to find a backup kit to deliver the glucagon to her daughter, but after the incident Bertron and her husband pledged to create a device that would make administering these drugs easier.

The trio's creation came in second in the Modern Marvels invention competition. The three are in talks with a medical-device company to manufacture it.

Boothe and his company, Corsair, EDA Inc., offer clients such services as initial-concept development, design selection and iteration, prototype fabrication, product validation, and low-volume manufacturing and production. This "fee for service" arrangement could prove a cost-effective way for someone to take a concept from the back of a napkin to a viable product. (www.corsair-eda.com)

Powered Ankle-Foot Prosthesis

In the winter of 1982, young rock climbing phenomenon Hugh Herr, age 17, and a friend attempted an epic climb to the summit of Mount Washington. Lost in a blizzard for three days without even a sleeping bag, their ordeal proved treacherous. Herr lost both legs below the knees to frostbite.

A few years after the terrible accident, Herr returned to the classroom to pursue a career in science so he could improve the lives of physically challenged people. He obtained an undergraduate degree in physics, a master's degree in mechanical engineering from MIT, and a PhD in biophysics from Harvard. He helped design his own prostheses and eventually climbed rock and ice walls at a more advanced level than before the accident using the special artificial feet he developed for vertical terrain. In 1989 Herr earned a place in the Sports Hall of Fame.

Recognized for his breakthrough innovations in prosthetics and orthotics, he was awarded the 2007 $250,000 Heinz Award for Technology, one of the largest achievement prizes in the world. Most recently, Herr and his biomechatronics research group at the MIT Media Lab have developed a robotic ankle-foot prosthesis capable of propelling the wearer forward and varying its stiffness over irregular terrain, successfully mimicking the action of a real biological ankle, and, for the first time, providing amputees with a truly humanlike gait. Hugh explains that the device mimics the elegance of nature, as a musclelike robotic assist releases three times the power of conventional prostheses to propel the body upward and forward in walking.

Handheld Medical Scanners Linked with Cell Phones

Although diagnosis and treatment of roughly one-fifth of all diseases can benefit from medical imaging, this is a procedure out of reach for millions of people in the third world because the equipment is too costly to maintain. Professor Boris Rubinsky and colleagues in the Bioengineering Department at UC Berkeley have invented a system to make imaging technology inexpensive and accessible for these underserved populations.

With Rubinsky's system, handheld medical scanners coupled with cell phones would allow medical scanners to aid people currently without access to ultrasound, X-rays, and other machines used to detect tumors or monitor fetuses. The simple portable scanner plugs into a cell phone, transmitting raw scanning data to processors which create images to relay back for doctors to view on the cell phone screen.

Bone, muscle, and diseased tissue such as tumors all conduct electricity differently. To spot these telltale differences, up to 256 electrodes are attached to the body and voltage is applied. The current between them is then measured, and the information is sent from the electrodes to the cell phone via a USB cable. From there the data is sent via text message to a remote server, where it is processed to create an image of the tissue. Differences in resistance, caused by tumors, show up as different colors. This is sent back to the doctor's cell phone for analysis. While the ultrasound scanner might cost $1,000, a whole ultrasound machine costs around $70,000, making this new technology a blessing to third world countries.

Environmentally Friendly Inventions

As the energy crunch worsens, inventors are finding it's a good time to be green. Growth in green or clean technology investment has skyrocketed and green U.S. technology companies will receive billions in backing in the coming years. Clean technology includes anything that uses energy efficiently.

Green Home Technology

Inventors continue to make big advances in incorporating green design into home technology.

SUCCESS STORY

Sykes's Enertia Building System

Inventor Mike Sykes won the 2007 Modern Marvels Invent Now Challenge, sponsored by the History Channel, Invent Now Inc., *Time* magazine, and Lexus. Sykes's Enertia Building System is a patented, environmentally friendly ideal house, refined and perfected over twenty-two years. The inventor uses two wooden shells for the home, one inside the other circulating air, and a passive solar design that creates a built-in "biosphere" in gradual but constant motion. Energy from the sun and geothermal stability from the ground create a temperate climate that buffers the primary living space, mimicking the greenhouse effect in the Earth's atmosphere. Sykes started building log homes to help pay his way through school. Today he sells kits of his patented energy-efficient homes on his Web site (www.enertia.com).

SUCCESS STORY

Green Grid Roofs

GreenGrid, a modular system of preplanted, recycled-plastic trays, presents a lightweight, low-cost alternative to permanent systems. The temperature of a flat, black roof can easily exceed 160 degrees Fahrenheit on a hot day—hot enough to fry an egg. The same roof covered in drought-tolerant plants will be as much as 100 degrees cooler—dramatically reducing the cost of air-conditioning. In many cases, plants can be installed atop existing roof gravel, residential or commercial. And they look so cool and green, too.

In Europe, green roofs have been around for a while, but they are costly to install and costly to maintain. The GreenGrid System offers a distinct advantage over traditional European-style green roofs due to its simplicity of design and cost competitiveness. The GreenGrid modular system, made of recycled materials, can be installed without specialized equipment. Installations, depending on roof size, can be completed in a few days to a few weeks. (See www.greengrid.com.)

Alternative Energy

Windbelt: Cheap, Efficient Wind Power

As a student at MIT, Shawn Frayne traveled to Haiti to help villagers. There, he saw a need for small-scale wind power that could economically replace kerosene lighting. He created the Windbelt, a way of tapping into wind energy without the construction of expensive giant turbines. He uses what is called aero-elastic flutter—most drastically exhibited in the Tacoma Narrows Bridge collapse—to create cheap electricity. Researchers at Humdinger LLC, the company pushing forward the Windbelt technology, have discovered that it can also be a powerful mechanism for catching wind beyond the reach of traditional turbines. Prototypes generated forty milliwatts in ten-mile-per-hour slivers of wind, making this device ten to thirty times as efficient as the best microturbines. A medium-size Windbelt building block has an initial target cost of $2 per rated watt. This is two to four times cheaper than photovoltaic systems and less than megawatt-size turbine wind harvesters.

A Windbelt can be made for under $5 and is easily and cheaply mass-produced, with little installation required. Frayne's invention was awarded the 2007 Breakthrough Award from *Popular Mechanics*.

Frayne started up Humdinger Wind LCC to market his device and supply Windbelt generator kits in a box. He hopes to fund third world distribution of his Windbelt, saying in a 2003 interview in the *New York Times*, "I learned in an economics class that if someone has a good idea and they can implement it in a third world country, they can dramatically change the economy of the country. I was surprised by how much technology can affect the well-being of a people."

SUNGRI: A New Solar Energy System

Ever use a lens to magnify sunlight and produce a very bright, hot spot? A new solar energy system from SUNGRI uses that simple principle and claims to be able to produce electricity for five cents per kilowatt hour.

The concentrated solar technology is called Xtreme Concentrated Photovoltaics and it works by using a magnifying glass to concentrate sunlight, making the light 1,600 times brighter. The system turns concentrated sunlight into electricity on a solar photovoltaic cell. The extreme heat of the sunlight is removed using SUNGRI's proprietary technology. As the process converts the sunlight directly to electricity without any intermediate step, it is not labor intensive to run or maintain. The technology will enable power companies, businesses, and residents to produce electricity from solar energy at a lower cost than ever before.

SUNGRI was formed by five inventive and experienced individuals with the goal of creating a renewable energy source available at fossil-fuel prices. The new solar energy system was announced by SUNGRI at the National Energy Marketers Association's 11th Annual Global Energy Forum.

Ocean Wave Energy Technology:
OceanLinx Ltd. and Pelamis Wave Power

Hawaii has the most powerful waves per square meter in the world, and in February 2008, OceanLinx Ltd., an Australia-based international high-tech company, announced plans to provide electricity to the island of Maui from Hawaii's first-wave energy.

The plan is to provide up to 2.7 megawatts from two to three floating platforms located one-half to three-quarters of a mile off the coast. The unique system combines the established science of the oscillating water column with OceanLinx's patented turbine technology. Rising and falling sea swells push and pull air past the turbine and its blades shift in response to the direction of the air flow so it turns continuously in one direction. Electricity generated is then brought ashore through an under-sea cable to a substation tied to the island's electrical grid.

On the island of Oahu, the Office of Naval Research is monitoring an experimental wave-energy buoy. The buoy has a continually bobbing motion that drives an electrical generator.

Pelamis Wave Power Ltd. is the manufacturer of a unique system designed to generate power from ocean waves, the first to be used in commercial wave-farm projects.

Off the coast of Portugal, the Agucadoura wave farm operates the world's first commercial power plant convert to the energy of sea waves into electricity. Three articulated steel "sea snakes" are moored to the seabed. The machines are positioned toward the waves so sections move with the waves. Each joint contains a hydraulic pump, which pumps high-pressure liquid through motors that in turn drive power generators. The energy is then transmitted to a substation on shore via subsea cables, generating enough megawatts to supply 1,500 households with electricity. The project will be expanded nearly tenfold over the next few years with some twenty-five "sea-snakes."(See www.pelamiswave.com.)

SUCCESS STORY

Eco-friendly Fridge

Chosen as one of *Time* magazine's best inventions of 2008, the eco-friendly fridge is based on an invention originally patented in 1930 by Einstein. It uses ammonia, butane, and water instead of Freon, a serious contributor to global warning. Apparently, Einstein's version wasn't all that effective, but scientists at Oxford University have tweaked his design. The fridge requires very little energy and the Oxford researchers believe it will eventually compete in the marketplace.

Safety-Oriented Inventions

Anecia Survival Capsule

Simon Cowell, host of *American Inventor*, referred to inventors as "normally normal people with normal jobs who have ended up sacrificing tons of money and a lot of time in the belief that this thing is going to turn them into a billionaire." However, the first contest winner invented his product to save children's lives.

In 2006, Janusz Liberkowski's Anecia Survival Capsule, a safer child's-car-seat, won the very first *American Inventor* prize of $1 million. Liberkowski was given the opportunity and the engineering resources of one of the biggest safety seat manufacturers, Evenflo, to continue to develop his product. In Liberkowski's car seat, the baby sits inside nested spheres that can spin and automatically position the child's neck and back so they are perpendicular to the impact force in a collision.

ImpactShield

In 1999, as Hurricane Irene headed for the Florida coast, residents scrambled to protect their homes with heavy, sharp-edged storm panels. Aspiring inventor and mechanical engineer Cameron Gunn decided that there must be a better way to protect homes and came up with his invention—ImpactShield.

ImpactShield is a fabric covering stretched as tight as a drumhead to protect a house's windows during a hurricane. Gunn's patented breakthrough includes stretching heavy polyethylene fabric taut with a crank so that projectiles bounce off. He sells his product online and for the past two years he has worked full-time to bring ImpactShield to market. He hopes to have his invention in a major home-improvement retailer soon. Making the top twenty-five of the History Channel's Modern Marvels Invent Now Challenge has brought attention to the product. Gunn sells his safety products at Shield Technology Group (www.shieldtechinc.com).

Functional Clothing: "Glitterati Clothing"—Nanofabrics That Repel Germs

Slinky. Sexy. And germ-repellent. Fiber scientists and fashion-design students at Cornell University have designed a two-toned gold dress that prevents colds and flu and a metallic denim jacket that destroys harmful gases and protects the wearer from smog and pollution.

The project began when Olivia Ong, a senior design major at Cornell, approached Juan Hinestroza, assistant professor of fiber science at Cornell, with what seemed at first a crazy idea. Ong said she was familiar with nanotechnology and wondered if it could be incorporated into her fashion line. Having lived in smoggy Los Angeles, Ong wondered if the science lab could produce a fabric that protected against smog and bacteria.

Hinestroza at first experimented for fun. He and postdoctoral researcher Hong Dong dipped positively charged cotton fabric into a solution of negatively charged silver metal ions. Silver possesses natural antibacterial qualities that are strengthened at the nanoscale, destroying bacteria and viruses and also reducing the need to wash the dress. Electrostatic forces bind the metal ions and cotton together, and the smaller the metal particles are, the greater the surface area for interactions with microbes or smog in the atmosphere. The denim jacket includes a hood, sleeves, and pockets with soft cotton embedded with palladium nanoparticles that act like tiny catalytic converters to break down harmful components of air pollution. Such properties would be useful to someone with allergies or to protect against harmful gases in a polluted city. To create the jacket material, Dong placed negatively charged palladium crystals onto positively charged cotton fibers.

These are among the first garments to qualify as nanotextiles, fabrics in which active nano particles are evenly distributed and less than 100 nanometers in diameter, or about one-thousandth the thickness of cotton fibers. The nanoparticles are so small that the clothing feels and drapes like a soft cotton T-shirt.

As a few yards of the treated fabric costs about $10,000, the germ-busting clothes aren't likely to be mass-produced soon; however, the clothing has attracted attention from clothing manufacturers and tech blogs. Hinestroza has been called in to brief the military on the project, as clothing that could protect against all kinds of poisons would be priceless during chemical or biological warfare.

Computer Technology

Apple Computers

In 1976, Steve Jobs and Steve Wozniak quit their day jobs at Atari and Hewlett-Packard, respectively, and founded a computer company in Jobs's parents' garage. They called the company Apple. Their goal was to create an inexpensive and easy to use personal computer. They were totally unprepared for their first commercial order from fellow students for fifty computers.

To raise the needed $1,300 for parts, Jobs sold his old VW bus and Wozniak sold his Hewlett-Packard calculator. On April Fool's Day in 1976, they introduced their first system, an encased circuit board known as the Apple I, which sold for $666.66 at the local electronics store. By 1977, Apple sales hit $800,000, and Apple went on to become a Fortune 500 company in a record five years!

VTV (Virtual TV)

Eric Gullichsen and Susan Wyshynski were among the talented young computer engineers and experts lured to Silicon Valley. Fascinated by the world of virtual reality, they wanted to get inside their own fantasies and experientially inhabit these worlds. Living on a houseboat and without much cash, they invented Virtual TV, or VTV.

The VTV computer program performs perspective corrections to a distorted image. Thanks to these inventors, the viewer can become immersed in a computer-generated real-world environment. Their program, which Wyshynski says is best described as "the sphere," drops viewers into the middle of a basketball game where, wearing virtual "i-glasses" they can look around, 360 degrees, in the video world. The program also is used in games like *Doom* to speed up turning and panning in the 3-D world, and to create virtual online real estate tours.

Wyshynski says, "Until then virtual reality had been limited to computer graphics, and complex production tools kept VR in the hands of engineers. VTV360 brought traditional film and TV into the VR realm and put production tools back into the hands of storytellers." VTV is suited to a variety of markets including computer gaming and the expanding location-based entertainment industry. It creates fast, high-resolution virtual experiences as it works with video and computer animation.

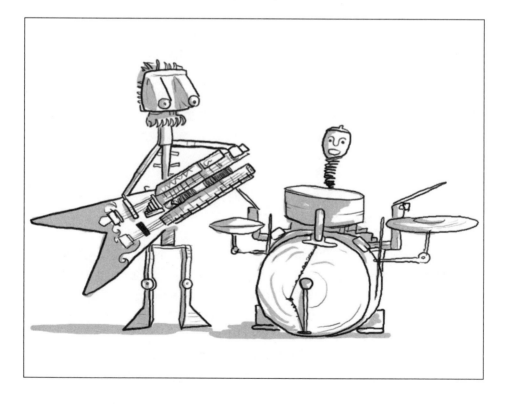

Robotic Bandmates

San Francisco rocker Jay Vance, fed up with losing band members to day jobs and girlfriends, upgraded his band with robots with scrap append-ages. He powered the mechanical rockers with a motherboard cobbled together from three computers and used pneumatic actuators to make robot fingers strum killer chords and drumsticks smack drum skins. Vance tours with his new version of a heavy metal band, Captured! By Robots.

Wi-Fi Detector T-Shirts: ThinkGeek

The inventive people at ThinkGeek have come up with T-shirts that are emblazoned on the front with a graphic of a radio tower. When the T-shirt wearer walks into any place with a Wi-Fi signal, the little radio tower illuminates. As the Wi-Fi signal increases, radio-signal symbols around the tower illuminate to a soft blue. The stronger the signal, the more the sym-bols light up in an outward direction. A strong signal means it's time for the T-shirt wearer to break out his portable PC and make the connection.

ThinkGeek.com started as a simple idea to create and sell products that would appeal to Web geeks and internet-technology buffs. Passion-ate about technology, three of the four founding members started an ISP (Internet Service Provider) in the Northern Virginia area in 1995. After a few years of the ISP gig, ThinkGeek launched their Web site on Friday the 13th, 1999. A month or so later they were so overwhelmed with orders that their system crashed. But all was not lost. Shortly after, ThinkGeek was acquired by Andover.net, which, after an acquisition and a couple of name changes, is now known as SourceForge Inc. So they are now part of cool sites like slashdot.org, sourceforge.net, linux.com, and freshmeat.net.

All the founding members are still with the company and they con-tinue to come up with great ideas, grow their product line, and expand their staff. ThinkGeek says that new products come about as a result of a group brainstorming sessions. Shane Peterman, one of the founders, says, "We're always trying to either look for or think of new ideas and products, so it's kind of a never-ending stream of suggestions around here."(See www.ThinkGeek.com.)

SUMMARY—THINGS TO DO

- Look at existing technology.
- Keep an eye on what today's inventors are inventing.
- Carry an idea book.
- Once you have a great idea, ask yourself if your idea is realistic, unique, and financially viable—and if it fills a niche in the market. Virtually every new entrepreneurial effort started in order to fill a niche, e.g., cell phones replacing hardwired phones and herbal teas replacing Ceylon black tea.
- Check existing patents. You've come up with a great new product. But did you come up with it first? See Chapter 11 for ideas for a quick check of existing patents.

Making Your Idea a Reality: Inventor's Logs, Prototypes, and Marketability Evaluations

I have been impressed with the urgency of doing. Knowing is not enough; we must apply. Being willing is not enough; we must do.

—*Leonardo da Vinci*

To invent, you need a good imagination and a pile of junk.

—*Thomas Edison*

AN INVENTOR'S LOG IS A RECORD OF YOUR INVENTIONS. An idea that didn't work out a few years ago may provide the sudden inspiration for a brand-new invention tomorrow. The log is also proof of what you invented and when you invented it. The log should be in a bound composition book, not a loose-leaf book, and entries should be dated. Entries should be written in ink or typed and sections signed and dated by a reliable witness at least once a week. Your inventor's log should contain:

- Your ideas.
- Possible problems to solve. Possible solutions.
- Details of your invention. Make a diagram whenever possible.
- Materials used and costs of parts.
- Steps taken.
- Store materials found.
- Successes and failures.
- Questions and problems.
- People consulted.
- Books consulted.
- Useful Web sites.

PROOF-OF-CONCEPT PROTOTYPES

Put together a crude proof-of-concept prototype early in your invention timeline, long before you seek a manufacturer. Prototypes are essential to prove your concept to yourself and to investors and customers. They help you see if your invention is realistic, before you commit too much time and money. Although you don't need a working prototype to file a patent, in many cases it is helpful to prove that you can create a product that will work as you describe. Keep your costs down. Keep in mind that all you are trying to do is make a proof-of-concept prototype to convince yourself the idea is possible. It needn't be a production piece.

Find Materials

The cheapest approach is to build the prototype yourself. Once you have a possible solution in mind, the next step is to get materials. The typical materials I use for problem solving and inventing are usually sourced from my attic or utility room. Note that you have to balance how much trash, junk, and materials you can keep with some semblance of neatness or you'll be ridiculed in your neighborhood for being a pack rat. Worse, your spouse will get sick of looking at your garbage collection and may ultimately leave you—especially if your invention has not yet been successful!

My other sources of material are trash piles, Wal-Mart, and Home Depot.

I'll give myself one hour in a Wal-Mart or Home Depot (if I can't find a well-stocked trash pile) and pretend that I must solve the problem before I leave the store, even if it's in the most rudimentary form. This strategy leads to strange encounters with salespeople, especially if I am asking them something bizarre like whether or not it is possible to cut a toilet seat in half and place some rubber tubing around it, or some other madness.

Of course when looking at "materials" from sourcing stores, keep in mind that you will likely destroy them to create something else. Therefore, look at potential parts of products that you can cut out or morph into something else. A hacksaw and tape are your best tools and always keep your mind open to the possibilities of creating something new with components extracted from other products.

During the prototyping phase, it is useful to ask yourself questions about the commercial viability of your invention: Does it have commercial appeal? How can you keep the manufacturing costs low? Can it be improved?

Evaluate Your Invention

Check out products in similar categories to your invention and evaluate your invention accordingly. How will your invention improve the lives of consumers? Evaluate your invention and see how many benefits you can identify and how many you could possibly add.

- Is your invention cheaper or more expensive to build and use than current products? Can you cut costs without cutting quality?
- If weight is important, is your invention lighter or heavier than the current products?
- If size is important, is your invention bigger or smaller than the current products? In many fields, e.g., computers and cell phones, small things are in demand.

- Is your invention safer and healthier than what exists? Volvo achieved success by stressing the safety of their vehicles. Can you be the niche supplier that focuses on being the leader in safety?
- Is your invention faster than current products? Does it save your customers time?
- Does using your product improve productivity and cut labor costs, e.g., robots replacing hand assembly by workers?
- Is your invention easier to use than conventional products? Many innovations are difficult to understand and operate. Is your product people friendly? Can your customers use it without complicated instructions? Will they grow to love it and depend on it? Are the functions of your product easy to recognize?
- Is your invention easier to repair than the competition?
- Rate your invention's novelty factor.
- Is your invention's appearance an improvement on existing products? Consider how Harley-Davidson offers a wild selection of colors and customized motorcycles and Apple computers are compact with a futuristic design.
- How does your invention rate, precision-wise?
- What is the noise factor compared to the competition?
- Market size and penetration: Can you identify a specific target market and use this as a focus for marketing? What is the size of the available market? How easy is it to penetrate the market? What does the target market want and why will they want your invention?
- Quality. What is your invention's anticipated life cycle and how does this compare to the competition?
- Is there an existing need for your invention?
- Does your invention require new production facilities or only a small change to an existing production facility?

MARKETABILITY: COMPANIES WHO WILL EVALUATE YOUR INVENTION

Wal-Mart Innovation Network (WIN)

It's often difficult for an inventor to determine the commercial potential of what seems to be a terrific idea. Wal-Mart, wishing to assist America's inventors, many of whom constantly approach them with new products, offers the Wal-Mart Innovation Network (WIN), an evaluation service giving inventors an opportunity to have their inventions evaluated and its flaws identified. A bonus is that products that score highly can be submitted directly to Wal-Mart buyers for review and possibly market testing in stores. The WIN evaluation can save an inventor a lot of time and money and the cost is nominal, around $175, much less than a patent search. Contact WIN at www.wini2.com, or by phone at (417) 836-5671.

WIN provides the inventor with a thirteen-page report, a letter containing specific comments, and a final recommendation from a chief evaluator, plus WIN's 277-page manual on evaluating potential new products.

Inventions submitted to WIN do not need to be patented as they are submitted under a confidential disclosure agreement.

A Note on Market Analysis vs. Marketability Evaluation

A "market analysis" should not be confused with a "marketability evaluation." A market analysis merely considers trends, consumption, users, and types of products in the market of your invention. Many unscrupulous "invention promoters" sell only a market analysis to inventors and not the marketability evaluation that is really needed. Michael Neustel of Neustel Law Offices Ltd. warns that although the market analysis has some useful information, it is of limited value to inventors since it does not tell them whether their invention is potentially marketable.

Below is an example from www.neustel.com of the significant differences between a "market analysis" and a "marketability evaluation" for a new toothbrush invention, SUPER BRUSH, that brushes teeth 25 percent faster than conventional toothbrushes.

MARKET ANALYSIS

- 10 million toothbrushes were sold in 1998.
- 4.4 billion people use toothbrushes.
- 2.3 billion people replace their toothbrush every year.
- Toothbrush sales are expected to top 12.5 million in 1999.

- The average price for a toothbrush sold in 1998 was $2.75.
- The average cost to manufacture a toothbrush in 1998 was $0.35.

MARKETABILITY EVALUATION

- Eighty-five percent of consumers do not care if they can brush their teeth 25 percent faster.
- There are over 250 different competitive products in the toothbrush industry.
- The SUPER BRUSH is heavier than conventional brushes so it is less desirable to consumers.
- The SUPER BRUSH requires extensive training by a professional to ensure proper usage.
- Consumers of toothbrushes do not change types of toothbrushes easily.

Neustel points out that if one were to only look at the market analysis, the SUPER BRUSH would appear to be a great invention, worth investing thousands of dollars in—exactly what some invention promoters want you to think. However, after considering the marketability evaluation, it is obvious that the SUPER BRUSH does not have much of a chance for making it in the toothbrush market. A smart inventor would not spend time or money on the SUPER BRUSH without further research. They would invest their time and money into their next invention.

Some unscrupulous invention promoters will charge you as much as $1,500 for a market analysis, *not* a marketability evaluation. This is particularly disturbing when you can do your own market analysis for free at most local libraries or by using Internet resources, e.g., the U.S. Census Bureau, Industry Research Desk, and *Hoover's* Online.

SUMMARY

- Keep an invention log.
- Build a proof-of-concept prototype.
- Keep your costs down.

- Search for materials from trash piles and Wal-Mart and Home Depot stores.
- Keep asking yourself questions about the functionality of your invention—can it be improved?
- Evaluate your invention and constantly look for areas you can improve.
- Ask customers what they need from your product and evaluate its marketability.

Build a Show Prototype

Hell, there are no rules here, we're trying to accomplish something.

—*Thomas Edison*

Opportunity is missed by most people because it's dressed in overalls and looks like work.

—*Thomas Edison*

ONCE YOUR INITIAL PROOF-OF-CONCEPT DEVICE IS constructed, get ready to build your next presentable prototype. Scour the Internet, the *Thomas Register* (which lists manufacturers and is available both in a library and online), and any other sources you can think of to get "real" sample components to build your next presentable prototype.

Get as many free samples as you can: Tell people that you are building a prototype and it may lead to interest in your invention and sales for your company. While working on a prototype for the RescueStreamer, I used many plastic samples from companies all over the country to try to get the right mix of materials to make it work. Ultimately, I located one of the materials through the *Thomas Register*. My licensee is still using this company as the main source of the plastic film.

The LIFE/FLOAT and DeSalinator both use highly specialized materials that were difficult to locate. Fortunately, the Internet was around by that time and I was able to research and track down materials to make the technologies possible. Note also that although you may get to a technical person at a specific company and they may not have the product you need, don't hesitate to ask them for advice on where they think you could locate it. People are usually eager to help you out, especially if you treat them as professionals and respect their opinions as experts.

When making a prototype of my PocketFloat, I bought many inflatable pool toys to do my initial prototype and testing. Slight modifications to the over-the-counter products led to confirmation of the concept.

FINDING EXPERT HELP

If your product requires assembly or engineering beyond your skills, check out one- or two-person machine shops, auto repair shops, tool-and-die shops, or tradesmen, such as welders, who can work with you. Network. Ask each source, even if they can't work with you, for a suggestion or recommendation. Make sure you obtain confidentiality agreements from any tradesmen who work on your product, or have different tradesmen work on different parts without disclosing the whole.

The industrial design departments of colleges, universities, and technical schools are good places for design help, and lists of industrial design firms can be found through the Industrial Designers Society of America.

You can use Internet sites like www.emachineshop.com to pick a material, draw a part, and have it made and delivered to your door. Tailor the prototype to fit the requirements of potential customers; the more operable the better. With solid objectives you should be able to get a good prototype made for a reasonable amount of money.

Innovations in software and some hardware often require even less capital investment. Inventors Chris Banker, Mike Cretella, Jeff DiMaria, Jamie Mitchell, and Jeff Tucker spent just $700 and three weeks to tackle the ubiquitous computer mouse and transform it into a snugly fitting ring that speeds a cursor around a screen. The prototype, designed and constructed as a student senior project at Worcester Polytechnic Institute in Massachusetts, looks like a kid's toy but works flawlessly.

Check the Web for sites like www.inventionhome.com that utilize state-of-the-art technology to provide virtual prototypes that can be run and viewed directly from your invention portfolio Web site. A virtual prototype is an interactive 3-D prototype that will demonstrate and educate prospective buyers on your invention's features, functionality, and benefits.

TIP

Tip from 3-D Prototype Design (www.3dprototype.com)

3-D Prototype Design Inc. provides customers with physical proto-types, using custom software and technology to turn drawings into a solid form very quickly. This gives inventors the edge they need to get their product to the marketplace in a fast, cost-efficient manner.

"Our customers range from large, well-known corporations to small one-person operations," says Annette Kalbhenn of 3-D. "We have built many different prototype parts over the years, from objects you use every-day to things you haven't even seen yet!"

When asked why he got into the rapid prototyping business, Tim Deutschmann, president of 3-D Prototype Design, says, "You can take pictures to show people, but to actually have a working scale model—for sales, that's amazing."

3-D Prototype gives the following advice to inventors considering rapid

WHAT IS RAPID PROTOTYPING?

Rapid Prototyping (RP) is the term used specifically for the process of building physical 3-D parts directly from CAD (computer-aided draw-ing) files. Regardless of the method used, RP creates a prototype (or part) through an additive, layer-by-layer process without the need for tooling or molds. By cutting out the traditional steps once needed to make a sample, RP is a significantly more rapid process.

Why Spend the Money on a Prototype?

As an inventor, you wear many hats. You are not only the inventor, but also a member of the product-development team, a manufacturer, and often the sales force. All of these roles benefit from having a physical pro-totype before you move into manufacturing.

prototyping:

What should I do now that I've come up with an idea?

Whether it's a brand-new invention or a modification to an existing item, rapid-prototyping technologies require a 3-D computer file format of your part so that your exact design can be built in a three-dimensional format. Rapid prototyping, as well as many final manufacturing technologies, requires specific math data (computer files) in a 3-D file so your design can be built in a three-dimensional format for prototyping and subsequently for manufacturing. Another benefit to having your idea designed three-dimensionally is the ease of modifying and making changes as the idea grows and develops.

Hiring a mechanical engineer or designer is a good starting point if you are serious about your invention. The engineers often know what will work best from a functionality standpoint as well as how to save money and what will be practical when it comes time to manufacture your idea.

Once you are at a stage where you feel comfortable with your invention on the computer in 3-D, then take it to the next level—have a prototype made.

RP Prototypes Will Provide You With:

- The opportunity to test for functionality, ensuring the design and fit are correct and also giving you the chance to work out the wrinkles and make changes while the costs are still relatively low.
- A physical model of your idea that can be used for sales meetings, presentations (i.e., investors and banks), and customer feedback prior to going to the expense of manufacturing.
- A packaging stand-in model to save time, while the real product is being produced.
- A model that will clearly communicate what your part should look like during the final stages of manufacturing to ensure accuracy (especially when being made overseas, where language barriers often occur and you can't be there yourself).

Often inventors find their original idea morphs during the design stages as they learn what works best in reality and what their potential customers like or don't like about the product.

HOW DO I MAKE A PROTOTYPE?

There are many companies that can provide you with a prototype. Rapid-prototyping service bureaus are companies that own rapid-prototyping technology and provide you with prototypes as you need them in the materials you require. If you find a good one, they will help you out along the way with suggestions on how to build the most economical and suitable prototype for your needs.

As a new inventor, you should be concerned about protecting your idea. All rapid-prototyping bureaus should have no problem signing a non-disclosure or confidentiality agreement, as their main business is building new or revised concepts.

Feel free to ask as many questions as you need to understand the processes and if the end product is right for you. If you have a concern or special requirement with your part specifically, be sure to mention it. If you want to see a sample of the material, most companies will send one once the quoting process has been completed so you can access the function and finish.

Whether it's a brand-new invention or a modification to an existing item, a 3-D CAD file of your design is needed to get a quote or build a part. Consult your rapid-prototyping bureau about what is needed. Ultimately, all forms of rapid prototyping will use an STL file, but the extensions .IGS and .STP are also acceptable, as they can be converted by your rapid-prototyping bureau into STL files. In some cases, a native file can be supplied and the conversion can then be made into an STL file.

From there, simply send your CAD file in the format requested to your rapid-prototyping bureau and they can give you a quote. If you approve it, your rapid-prototyping bureau can build from the same file you supplied and courier it to you once complete. With the convenience of e-mail and the Internet, you never need to personally visit the facilities.

Wait Time

This can vary from twenty-four hours to five to six days (sometimes more) depending on quantity, size, machine availability, and money-saving suggestions.

Types of Prototypes

What type of prototype do you need? Not all prototypes are created equal. There are four basic prototypes available:

Selective Laser Sintering (SLS)

SLS prototyping sinters (or melts) one layer of powder material (such as nylon) to the next, creating a solid plastic component. This process can also create rubberlike parts and metal prototypes.

ADVANTAGES:

- Fully functional, durable prototypes.
- Nylon material parts can be tapped, drilled, snap fitted, assembled, and painted.
- Living hinges and functioning springs can be produced.
- One or multiple pieces can be rapidly produced.

DISADVANTAGES:

- As the material is very strong, it's a little more difficult to sand and finish the surface.
- Not available in translucent material.

SLA (Stereolithography)

Stereolithography (SLA) creates prototypes layer by layer using an ultraviolet laser, a vat of photo-curable liquid resin, and a controlling system.

ADVANTAGES:

- Ability to produce translucent parts
- Can capture fine detail
- Resolution as low as .001-inch layers (with some machines and materials)
- Easier to sand and smooth surfaces

DISADVANTAGES:

- Brittle materials that can break and fracture with normal handling
- Parts continue to cure once completed, over time making them more fragile.
- Not fully functional prototypes

Fused Deposition Modeling (FDM)

The FDM process builds prototypes up layer by layer in material such as ABS, polycarbonates, or wax using extruding heads.

ADVANTAGES:

- Can quickly build one or a few pieces at a time
- More durable than an SLA or 3-D printer prototype
- Has some functionality due to nature of ABS material

DISADVANTAGES:

- Rough surface finish (you can easily see the extrusion layers and places where supports once existed.)
- Only one or a few built at a time
- Detail is difficult to capture due to the nature of the process.

3-D Printing—Z-Corp

This method is termed 3-D printing because it is based on the same technology that a common 2-D ink-jet printer functions on. Z-Corp uses a powder composite material, similar to plaster, and a liquid binder (adhesive) that is extruded from heads, very similar to ink-jet heads but with adhesive, instead of ink, in the cartridges. The adhesive can also "print" in color.

Similar to SLS, the CAD file is built layer by layer in a powder material, but the final products are very different.

ADVANTAGES:

- Economical parts can be created.
- Can produce color parts, like a topographical map
- Fast way to produce RP parts

DISADVANTAGES:

- Not a functional part
- Grainy in finish (similar to a sugar cube) and brittle
- Limited ability to capture detail

Using the Correct Terminology—SLA vs. Rapid Prototyping

Many mistakenly use the term SLA (Stereolithography) when they need a rapid-prototype part produced. Rapid prototyping (RP) is the broad term for the various processes that are available. Understanding the terminology will help you to better determine what kind of final RP part you want as well as compare quotes more effectively.

What Does an RP Prototype Cost?

The cost of your prototype will depend on the physical size, area taken up, material and method used, and quantity of your part. Depending on the above criteria, at times you can have twenty-five pieces made for $250 and at other times one part can cost $5,000. Rapid prototyping is a sophisticated technology that can easily provide you with an accurate quote with your CAD file. Even if your idea is not completely fleshed out, you can get a good estimate on pricing from your preliminary file.

Prototyping will likely cost you only a very small fraction of the costs associated with manufacturing and taking your product to market. Before you take that next big step, a prototype is a sound investment.

Main Checkpoints for Effective Quote Comparisons:

- Are the RP processes being compared the same process (e.g., SLS to SLS)?
- Are the parts being built at the same layer of thickness (e.g., .004 in. to .004 in.)?
- Are they using the same material?
- Are they quoting in the same currency?
- What are the shipping costs? (You need all your numbers.)
- Timing: Can they deliver to meet your needs?

Who Uses Rapid Prototyping?

Many inventors, including large corporations, who are creating, manufacturing, or selling a physical product, incorporate rapid prototyping into their development process. 3-D prototype design has produced thousands of prototypes in design programs for large automotive and toy corporations, as well as parts for small companies and individual inventors.

Product Development Companies

Product development companies are another solution to an inventor's prototype needs. These companies provide full-service assistance in prototyping, marketing, financing, legal protection, manufacturing, and packaging. The product development firm will work toward building a prototype that can be affordably manufactured. To do this, they might want to know how many units you plan to make in your initial manufacturing run and what you want the retail price to be. Using this information, they will have a prototype built out of a material that simulates as closely as possible the eventual materials used for manufacturing.

One such product development company is T2 Design Corp. in Santa Monica, California.

TIP

Tip from Paul Berman, T2 Design Corp. (www.t2design.com)

Paul Berman has twenty-six years of engineering and mechanical design experience, including fifteen years of consumer product design. Berman and the members of his company, T2 Design, perform product and invention evaluation, conduct patent searches, and have over a decade of experience in conceptual design, package engineering, and prototype construction.

Berman explains, "We evaluate and research the idea, including patent searching, before we move on to the essential stage of designing a prototype. The raw idea is styled and designed into a functioning prototype, including mechanisms or electronics if needed to make it viable."

Berman has developed over 125 inventions and has had over 15 patents issued, and his experience includes twelve years of computer-aided aircraft and space-structure design and integration. He worked on the research, detailed drawings, stress analysis, selection of materials, and conceptual design of the B-2 stealth bomber, NASA's *Galileo*, and the *Jupiter* space probe, among many other aircraft and space vehicles. Among a diverse range of inventions, T2 Design developed Skids (inline front skate brake) and andirline headphones (pneumatic transmission type), just to name a couple.

Berman advises that many potential investors will want to see a prototype before any money changes hands. And there is another equally significant reason for developing a prototype: You have to make sure your product not only looks the way you envisioned, but works the way you want it to. Many inventors are shocked to discover that what works on paper doesn't always work in practice.

Inventors who approach T2 Design first go through a one-hour consultation during which the product is discussed and future actions are identified. This costs $125. "We can shed a little light on what we think the odds of success for the product are," says Berman, who estimates at least half the ideas brought to T2 Design are rejected at this stage.

Generally, the next step is a patent search. T2 charges $450 for a manual patent search that includes a patentability opinion from a patent attorney. This means an individual actually goes to the U.S. Patent and Trademark Office in Washington, D.C., and hand searches through the patent repository library. When the search results are in, T2 contacts a patent attorney, who will review the results and provide a written opinion on the patentability of the idea. Market research and the building of an engineering model typically precede the prototype.

T2 Design can assist inventors after the prototype stage too. "Conceivably," Berman says, "we can take a product all the way through a license deal."

What Does a Product Development Company Look for in an Invention?

Berman cites eight key elements of a successful product:
- It is uniquely patentable.
- It can be manufactured with technology already in existence.
- Research and development costs are not too extravagant.
- There is a demand for it.
- The retail price ensures profit at each distribution level.
- Tooling costs are within reason.
- It isn't too complex, so consumers can use it easily.
- There aren't too many obstacles to market entry.

How Do You Find a Product Development Company?

Ask for recommendations from friends, family members, and personal contacts. Ask local inventors' organizations and small-business organizations. If all else fails, look in the Yellow Pages. Or call small companies with products similar to your own idea, contact the inventor, and ask if he used the services of a product development company.

PROTOTYPING, DESIGNING, AND TESTING YOUR IDEA

A commercially successful invention must be well designed. It must be more than just a good idea. Its design should minimize manufacturing costs, be eye catching and functional, and work to maximize durability. Your ability to license an invention on favorable terms is enhanced when you offer a manufacturer or marketer everything they need to make a prompt and appropriate decision. Berman recommends looking for an experienced product development team in a prototype company that offers to do the following:

- Design and build an attractive, working prototype.
- Design good packaging.
- Obtain the appropriate legal protection.
- Produce engineering drawings.
- Generate manufacturing-pricing data.
- Line up factory production facilities.
- Secure positive feedback from potential store buyers who are ready to order the product.

What Will It Cost to Prototype Your Invention?

At T2 Design, a typical consumer-product idea costs between $1,000 and $8,000 to develop. Complex items, particularly those with custom-designed electronics, are more expensive to develop. Important factors that determine the cost are mechanical complexity, the need for customized parts, and molding requirements. For prototypes with simple electronics, the starting point is approximately $3,500. Prototypes with sophisticated electronics usually start at around $7,000.

Prototypes That Came Through

The Harbor Wing Autonomous Unmanned Surface Vehicle (AUSV)

An avid sailor and ocean racer, Mark Ott spent a lot of time racing his self-built thirty-by-thirty foot trimaran in the Hawaiian Islands. A constant inventor and engineer, Ott always came up with ways to improve the trimaran's design, leading to the idea of a safer, faster vessel, powered predominantly by forces of nature, with a computer-controlled sail. But Ott, the founder and manager of a Rolls Royce repair facility in Honolulu, Hawaii, needed expert help with the design and the building of a working prototype that would turn his idea into a reality.

Using the Internet, Ott located world-class experts in design engineering, wingsails, boat platforms, guided missile design, hydrofoils, and aerospace composites, and designers of C-Cats for America Cup syndicates. He called cold, saying "I have this idea. You're good at doing a certain piece of this. Let's get together."

At a meeting in Connecticut, the world's leading authorities on platforms and hulls got together at an engineer's home and talked about the concept. Would it work? They brainstormed for ages and then agreed—yes, it would!

"Good idea," the experts said. "But it will take millions. That's your job."

On the plane flying back to his home in Hawaii, Ott pondered what a huge effort it would take to find the money to design and build the necessary prototypes. Moving from the visionary side, he started to pursue the business side. How would he find the money?

As he specialized in Rolls Royce repair, he knew a number of wealthy people. He asked advice. How would they go about doing this? This led to a business contact, who knew a scientist who had worked with Jacques Cousteau and knew an admiral in the United States Navy. Admiral Stuart Platt was interested in the concept. His input shifted the focus from a vehicle for recreational purposes to a fully autonomous unmanned surface vehicle (AUSV) that could benefit the U.S. Navy. He drew on his extensive commercial and military experience to provide leadership and direction to a new company, Harbor Wing Technologies Inc.

The project still had a long way to go. At one stage, Ott remembers, money was so tight that he and his wife gave up getting the morning newspaper. But military funding for work on the project and the prototypes came through. The essence of the AUSV's success is its advanced, innovative technology. To integrate such concepts and products into its design, Ott and Harbor Wing Technologies reached out to a huge number of specialists, including experts in wingsail design, hydrofoil design, aeronautical design, and robotics and embedded systems.

The high-tech AUSV has fulfilled all expectations. Tests of the prototype show that it meets fundamental military and government requirements for reconnaissance and surveillance, drug interdiction and search and rescue, environmental-sanctuaries enforcement, range operations, and safety monitoring and mine-survey countermeasures. Its commercial applications are industrial assets and shipping protection, undersea oil and gas exploration, ocean surveying and mapping, fisheries support, marine-mammal monitoring, and recreational boating.

Using a wingsail that can rotate a full 360 degrees to let the vessel maneuver efficiently upwind or downwind, and a custom-made guidance system that can relay vital navigational and situational data to a semiportable command station able to pilot the vessel if needed, Harbor Wing Technologies is presently also developing a luxury recreational catamaran that will bring no-fuel sailing to a larger market. (See www.harborwingtech.com.)

Consider Getting a Professional Drawing

If your prototype proves too difficult to build, consider simply paying a draftsman to make a professional drawing of the invention. A picture says a thousand words. For the RescueStreamer, I actually invented the technology based on my vision of "before" and "after" pictures with a person disappearing into the big blue ocean without a streamer ("before") compared with the person trailing a nice, long orange tail that made him look like an amoeba or a person with an exclamation point attached to his butt ("after"). It was a basic and primitive approach, however it worked at getting the point across and it still works.

To get the RescueSteamer "before" and "after" pictures, I persuaded one of my macho North Shore surf buddies to be the guy lost at sea. We headed out to sea in a boat with a photographer hovering overhead in a helicopter. As I had to get the boat out of the picture, I left him alone without the streamer in the deep ocean. He was as white as a ghost when I came back to give him the streamer for the "after" photo. Without the streamer, I almost lost him in the ocean.

I have sold many inventions with nothing more than a well-thought-out drawing. The PocketFloat, LIFE/FLOAT, DeSalinator, Portable Motion Detector, and vSAR were all awarded military development grants based on sketches alone. (On a side note, the patent office only accepts drawings—it is a common misconception that you have to physically show up at the patent office with a physical model or prototype of your invention.)

Test Your Product

Products need to be thoroughly tested before taking them to the marketplace. Keep an open mind for change. You may even discover a better way of completing a certain function. If difficulties arise, try different things until you overcome the problem and, if necessary, get expert advice. The following success stories illustrate how important it is to keep testing to prove your product works on all fronts.

SUCCESS STORY

Zubbles

Tim Kehoe wanted to make brightly colored bubbles. Bubbles have been a popular product, purchased in large quantities for generations. Remember blowing bubbles! Kehoe mixed batch after batch of brightly colored solutions, but the dye would run down the sides and into the bottom of the bubble.

Then suddenly, after two years of experiments, a brightly colored bubble floated across the kitchen. Kehoe thought he'd make a fortune.

He started showing his bubble to toy executives, who were excited too, realizing what a huge profit potential the bubbles had. But then the bubble burst. And it left stains throughout the boardroom.

Kehoe worked for eight years and eventually produced a bubble that didn't stain clothes and furniture. Excited investors rented a bubble machine, and children and their parents chased colored bubbles gleefully until they realized that the new bubbles stained their skin. They weren't impressed.

Then Kehoe did something he probably should have done earlier: He consulted an expert with a PhD in dye chemistry. After many further experiments with dyes, the expert came up with a dye that disappears within thirty minutes, leaving no stain. Kehoe's advice: Don't give up. And if you need expert help, consult an expert.

Popular Science named Zubbles the Innovation of the Year for 2005, and *Readers Digest* named Zubbles one of the best innovations of 2006. Kehoe, through his company Ascadia Inc., strives to produce products that fill people with wonder—no small task—and as of 2008 they are still working to bring Zubbles to the marketplace.

Portable Water Filter

Inventor Dean Kamen, an immensely successful inventor and the man behind the Segway, goes all out to prove his technology. To prove that his portable water filter really works, Kamen drank a glass of his own pee at a 2004 Medical and Information Technology conference.

Kamen spent several million dollars and three years of his life developing the filter that purifies any type of tainted liquid and turns it into pure, distilled water. And now, even Bono believes. Kamen's intent is that the water filters be used to tackle third-world water shortages.

Nutty Buddy Protective Cup

Mark Littell, former major league pitcher, also goes absolutely all out to prove his product works. Showing his Nutty Buddy protective cup in a YouTube video, Littell risks his manhood by having a machine fire a baseball that smacks him right in the you-know-what with a thud.

Littell spent $40,000 on developing and patenting the product, making his first prototype from moldable plastic and two golf balls.

Presently, Littell sells his product through the Web. He challenges other cup company CEOs to stand by their product like he has, saying, "You put your cup on, and I'll put my cup on, and we'll see who is left standing."

SUMMARY

- Consider the best way to build a show prototype.
- Work the Internet, 1-800 numbers, and the *Thomas Register* for sample components.
- Look for free samples.
- Check machine shops, auto repair shops, tool-and-die shops, and tradesmen who can work with you.
- Obtain confidentiality agreements.
- Check industrial design departments of colleges, universities, technical schools, and design companies for design help.
- Check Internet sites like www.emachineshop.com.
- Check www.inventionhome.com for state-of-the-art technology providing virtual prototypes.
- Consider getting a professional drawing from a draftsman.
- Consider using a rapid-prototyping bureau.
- Consider the services of a product-design company.
- Test and prove your product.

Name Your Invention

*German engineer Konrad Zuse may have invented the computer, but he did not
have the foresight to name it after himself or I'd be typing on a LapZuse right now.*
— *Paige Wiser*, Chicago Sun-Times

ONCE YOU HAVE A VIABLE SHOW PROTOTYPE, IT'S TIME TO
choose a final name for your invention. This is almost on par with naming
your firstborn. But at least the choice is up to you, and you don't need to
name it after anyone's grandparents.

POINTS TO CONSIDER

- Search existing trademarks or potential names in the same field.
  What is the competition named? Make sure no one is using the
  name. Check domain names as you will want your company name to
  have a company.com–style domain name.
- Search the Internet to check.
- Check company names in the state where you plan to incorporate.
- Brainstorm. How do you want people to feel when they hear the
  name?
- What names convey your product's benefits?
- Why is your product unique?

- What lingo or expressions in your industry best describe the product?
- Choose a name that is easy to remember.
- Choose a name that is easy to pronounce and spell. Did you ever see a late-night movie about a band called "The Oneders" who had one big hit song, "That Thing You Do"? They had actually intended the pronunciation of their band's name to be "The Wonders" but disc jockeys mispronounced the name and called them the O-need-ers.
- Look for synonyms on Thesaurus.com or in a good hard-copy thesaurus.
- Combine words.
- Use puns.
- Use rhymes. Check out RhymeZone.com.
- Experiment. Play around with combinations of various words and partial words. Make a list of favorites.

ONCE YOU HAVE THE NAME

- Stake your claim. Register your desired name or file your incorporation papers immediately.
- Start using either TM (trademark) or SM (service mark). You do not have to register these marks to use them.
- Get the domains. Find an inexpensive registrar and register your domain and any obvious variations. This will cost you around $10 a year per name and prevents poachers.
- Protect your brand. A U.S. trademark or service mark costs $325.

Avoid choosing a name that is so good that it is misleading and makes too many claims. The FTC (Federal Trade Commision) recently issued a ruling on the "Ab Force" device, stating: "A product name can help an advertiser convey a claim about the central attributes of a product. The product name 'Ab Force' . . . suggests that consumers will achieve more forceful or well-developed abdominal muscles. . . . The ads played an obvious role in conveying [the allegedly deceptive] implied claims to consumers."

The commission also found fault with "Smoke Away," a dietary supplement. Other products in trouble were "Fat Trapper" and "Exercise in a Bottle."

GOOD NAMES

Great names that convey the product's benefits include Kleenex and Plax Mouthwash. Market research shows that people like names with matching letters, like Coca-Cola.

NOT-SO-GOOD NAMES

Epoxyshield Concrete Stain Kits. The name is confusing. What does the product do? At first, I thought it must be a stain remover. Apparently, it's a good product that stains concrete. Would you eat Patagonian toothfish if it hadn't been renamed Chilean seabass (even though it doesn't belong to the bass family of fish)? Then there are products reviewed as "great products, bad names," including TomTom One Europe GPS Receiver, Coochy Shave Cream, and Garbage Energy Drinks. Check out the Name Inspector (www.thenameinspector.com) for some fascinating reading and an abundance of information on names.

The categories below are from the Name Inspector and are based on the morphological structure of names: what kinds of meaningful pieces they have and how the pieces fit together. They're listed in descending order of frequency.

Real Words

Names that are simply repurposed words. Such names can't be generically descriptive, because then they wouldn't be protectable trademarks, so they usually work through metaphor or metonymy (indirect association).

Pros: These names are short and come ready-made with rich, often multiple associations.

Cons: Expect to pay money—possibly a lot—to secure the URL. Trademarking can be tricky too.

Examples include Adobe, Amazon, Apple, Dapper, Ether, Expo, Flock, Fox, Grouper, Indeed, Inform.com, Live.com, Multiply, Pandora, Pluck, Revver, Riffs, Shadows, Sphere, Wink, Yahoo!, and Yelp.

Misspelled Words

These are simply words that have been misspelled to make them more distinctive. This addresses the URL/trademark issue.

Examples include del.icio.us (delicious), Digg (dig), flickr (flicker), Google (googol), Goowy (gooey or GUI), Snocap (snow cap), SoonR (sooner), Topix (topics), and Zooomr (zoomer). Foreign words: Renkoo (Japanese renku, a type of poetry), Rojo (Spanish "red"), Vox (Latin "voice")

Compounds

Each of these names consists of two words put together, with the first word receiving the main emphasis in pronunciation. (It doesn't matter if there's a space between words.) In most cases both words are nouns. Names with verbs in the second position are Bubbleshare, Google Talk, and possibly Tailrank (share, talk, and rank can all be nouns, but they're verbs under the most natural interpretation). Names with non-nouns in the first position are BlueDot, SocialText, JotSpot, Measure Map, and possibly Jumpcut, Rapleaf, and SearchFox. Again, the first words here can all be nouns, but they're more naturally treated as two adjectives (blue and social) and a bunch of verbs.

Compounds are a simple way to create new words and are very common in English (and other Germanic languages), so it's not surprising to find them high on the list.

Pros: The practically limitless number of possible combinations makes it easy to create a unique name. Interesting meanings can be created through the combination of words.

Cons: There are no huge drawbacks, which is one reason that compounds are popular, but they are longer than many other kinds of names.

Other examples include Attention Trust, Bloglines, Facebook, Feed-Burner, Filmloop, Firefox, Netvibes, Newsgator, OPML Editor, Pageflakes,

Photobucket, Powerset, Salesforce, Songbird, TagJag, Tagworld, TechMeme, Webshots, Wordpress, Video Egg, and YouTube.

Phrases

These are names that follow normal rules for putting words together to make phrases (other than compounds).

Pros: They sound linguistically natural and have clear meanings because they follow regular rules.

Cons: Phrase names can be long, and they can also sound awkward when used as nouns if they are not already noun phrases (e.g., "Have you tried iLike?").

Examples include 37 Signals, Adaptive Path, AllofMP3, AllPeers, Amie Street (could be a compound, but __ Street is such a common pattern), CollectiveX, iLike, Last.fm, LinkedIn, MyBlogLog, MySpace, PayPerPost, Planet Web 2.0, rawsugar, SecondLife, SimplyHired, SixApart, Stumble-Upon, TheVeniceProject (could be a compound, but the "the" makes it phraselike), and TopTenSources.

Included in this category are names that consist of a company name or prominent brand name followed by a generic noun. In these names, the first word functions as a kind of modifier of the second.

For example, AIM Pages, Google Reader, Google Video, Microsoft Expo, and Yahoo! Answers.

Notice Google Talk is not here. That's because Google Talk is pronounced with the emphasis on Google, which means that the whole thing is treated as one word. As far as the Name Inspector knows, all the names immediately above are pronounced with some emphasis on each word, and the main emphasis on the second. Does anyone disagree?

Blends

Each of these names has two parts, at least one of which is a recognizable portion of a word rather than a whole word.

Pros: When they work, blends can be short and elegant and have all the advantages of compounds.

Cons: When they don't work, blends can be awkward and/or have obscure meanings.

Examples include Maxthon (max + marathon), Microsoft (microcomputer + software), Netscape (net + landscape), Newroo (new + kangaroo), PubSub (publish + subscribe), Rebtel (rebel + telephone), Rollyo (roll + your own), Sharpcast (sharp + broadcast), Skype (sky + peer-to-peer), Technorati (technology + literati), Wikipedia (wiki + encyclopedia), and Zillow (zillions + pillow, with overlap of "ill").

Tweaked words

Some names are just words that have been slightly changed in pronunciation and spelling—usually with a letter replaced or added.

Pros: As long as people recognize the word, you get all its rich meaning while still having a distinctive name.

Cons: People might not recognize the word, and some of these names can be a little gimmicky.

Examples include Attensa (attention), CNet (It might stand for computer network, but who thinks of it that way?), eBay, edgeio, eSnips, iPhone, iTunes, Wikia, Zoho (Soho), Zune (tune), and Zvents (events).

Affixed words

These are novel forms consisting of a real word and a real prefix or suffix. Notice how common the -ster suffix is.

Pros: These names can be distinctive and meaningful while remaining relatively short.

Cons: Sometimes these names sound contrived. The meanings added by affixes are limited in variety and usually abstract (which means not very vivid).

Examples include Browster, CoComment, Dogster, Feedster, Findory, Friendster, Napster, Omnidrive, Performancing (performance isn't a verb, so it doesn't normally take an "-ing" ending), and PostSecret (post can also be a noun or a verb, making this a compound).

Made-Up or Obscure Origin

These are short names that are either made up or whose origins are so obscure that they might as well be made-up.

Pros: Made-up names can be short, cute, and very distinctive (and therefore easy to trademark).

Cons: Made-up names don't provide much ready-made meaning to work with (all the meaning has to come from sound symbolism). Good ones are hard to think of, and when they're short the URLs are likely to be taken

Examples include Bebo, Meebo, Odeo, Ookles, Plaxo, Qumana, Simpy, and Zimbra (taken from a Talking Heads song based on a nonsense Dada poem).

Puns

These names are words or phrases that have been modified slightly to evoke an appropriate second meaning. They're similar to blends, but they involve a coincidental similarity between part of the main word and the second evoked word.

Pros: Pun names can be fun and memorable.

Cons: Nothing sounds dumber than a bad pun.

Examples include Automattic (automatic, mat is Matt, the guy who started the company), Consumating (consummating, consumm is consum[e]), Farecast (forecast, fore is fare), Memeorandum (memorandum, mem is meme), Meetro (metro, met is meet), Meevee (teevee/TV, tee is me[e], the pronoun), and Writely (rightly, right is write).

People's Names (Real or Fictitious)

Some names are either pitched or recognizable as people's names. If the audience for a name doesn't see the connection, the name is just like a made-up one.

Pros: These names are short and give personality to a company (or product or service).

Cons: Aside from personality, these names don't provide meaning to work with. As with made-up names, good, short ones might not be available as URLs.

Examples include Bix (e.g. Bix Beiderbecke), Jajah (F. Jajah Watamba seems to be their fictitious spokesperson), Kiko (a name in Japanese and other languages), Ning (a Chinese name), and Riya (the name of a founder's daughter).

Initials and Acronyms

These are names made up of the first letter of each word in a much longer phrase name. Sometimes the letters are pronounced individually, in which case we can just think of them as initials, and sometimes the combination of letters is pronounced as a word, in which case it's an acronym.

Pros: These names provide short mnemonics for long, descriptive phrases.

Cons: Zzzzzz. Also, sometimes initials are short when written but long when spoken. For example, the initials www have nine syllables when spoken, while the phrase World Wide Web has three.

Examples include AOL (America Online), FIM (Fox Interactive Media), and Guba (Gigantic Usenet Binaries Archive).

The Name Inspector hopes that these name categories will be useful to people struggling with their own naming problems. They might suggest naming strategies or spur name ideas that wouldn't otherwise come up. Good luck in your naming endeavors!

SUCCESS STORY

How Starbucks Got Its Name

The founders of Starbucks, two school teachers and a writer, chose the name Starbucks in honor of Starbuck, the coffee-loving first mate in Herman Melville's *Moby Dick*. They thought the name evoked the romance of the high seas and the seafaring tradition of the early coffee traders.

The company's original logo, designed by an artist friend, was a two-tailed mermaid encircled by the store's name.

For a while, the founders changed the name to The Il Giornale Coffee Company and featured Italian décor and incessant opera music. But fortunately for coffee drinkers everywhere, eventually the original name of Starbucks prevailed. (Imagine trying to tell a hot date to meet you at a coffee shop whose name you couldn't even pronounce!)

ForceFins: Great Name and Great Design

When Bob Evans designed his Force Fins his inspirations were nature and ocean creatures. His notebooks are filled with photos and drawings of different kinds of fins, some like the fins of a squid, some like a tuna, some like a porpoise. He studied the fins of the fastest fish and tried numerous designs. He also extensively researched the movements of the human leg and of a diver's foot, including the laws of nature and physics applicable to his design.

For Bob Evans and his wife and partner, Susanne Chess, perfecting the fins and bringing the product to market was a long and arduous task. For two and a half years Evans worked ten hours a day on fin prototypes, not even doing his laundry—he says he just kept buying new swim trunks to wear. His drive paid off. After twenty-four tumultuous years, Force Fins have become a major market player in the swimming community and, finally, the dive industry. Force Fins are different from any prior fins, and Evans's work has been an inspiration to designers of the keel for one of the America's Cup teams. He has even been asked to apply his fin technology to increase the speed and efficiency of oceangoing ships.

The name Force Fin reflects the fact that, unlike other fins, these generate so much lift that they actually force themselves onto the diver's feet and do not need a strap to keep them on. While other fins act like an anchor and pull the diver down, these lift the diver's feet up and hold them up. Force Fins are half the size of other fins and divers using them expend half the energy to travel the same distance and carry the same load. The company is now selling winglets, called Force Wings, as an alternative attachment for any fin so that anyone can add some force to their fins (www.forcefin.com).

Checklist: Is Your Idea Feasible, Marketable, and Financially Viable?

Never waste time inventing things that people would not want to buy.
                                                          —Thomas Edison

Mr. Bell, after careful consideration of your invention, while it is a very interesting novelty, we have come to the conclusion that it has no commercial possibilities.

                    —J.P. Morgan, rejecting Alexander Bell's invention of the telephone

AFTER YOU HAVE EITHER A PROTOTYPE OR A PROFESSIONAL drawing of a prototype, you need to test your product personally. What do you really feel about it? Does it pass your own "wow" test—is it something you would want to have and be willing to buy?

Next, test it with friends and family. Does it score a "wow" reaction? Sharpen your presentation and listening skills. Take constructive criticism and incorporate it into future prototypes and designs.

Watch the body language of people as they critique your idea—if their eyes and faces light up as they talk to you and their body is turned toward you and receptive, they are probably truly impressed. If their arms are crossed and their legs and hips are turned to the door to get away from you, you're probably not making much impact! Sometimes people will say nice things; however, their facial expressions or body language contradict their statements.

Next, test with neutral third parties, and gather input along the way. After all the input from yourself, friends, family, and third parties, do you still think it is a good idea? Anyone make nasty comments? Someone once publicly called my PocketFloat "glorified water wings." That one hurt, but I had to roll with the punches. The other line I constantly get is "That is so simple, why didn't I think of that?"—which I take as a compliment, because the simpler and the better, the more elegant in my mind, as it adds up to less moving parts, less to break, and a more reliable product.

Learn your customers' requirements. Often customers buy products to get jobs done. Does your product help get the job done faster, more smoothly, and less expensively? Rework your ideas. Perform a marketability evaluation.

If after all of this, you are still positive about the idea and you can make it profitably, then it's time to protect it.

Summary:
- Test your product personally—does it pass your "wow!" test? Would you buy it?
- Test with family and friends.
- Test with neutral third parties.
- Learn your customers' requirements.
- Rework your idea.

PROTECT

Safeguard Your Intellectual Property—Paranoia Can Be Good

I will discuss the importance of education in the process of invention and believe invention should be taught in schools. Knowledge of patents and methods needed to protect originality are vital.

—Trevor Baylis (inventor of the Clockwork Radio)

IF YOU ARE CONSIDERING GOING INTO BUSINESS WITH your invention, you are entering shark-filled waters. These are the big boys, and they don't fool around. The only way to stay in the game as a little guy is by protecting your intellectual property (IP).

TRADEMARKS, DOMAINS, COPYRIGHTS, AND PATENTS

Utility patents are one of the best ways to protect your invention, however they are hard to get and very costly. Do your research before you put down $5,000 or more to formally apply for a patent through a patent attorney.

There are a couple of much less expensive routes in the world of intellectual property (IP) that you can pursue first. The least expensive is securing a Web site domain name.

Domain Name

It costs less than $10 per year to own a domain name. There are many services that help you find and secure a domain name (e.g., www.buydomains.com and www.godaddy.com). The shorter and simpler the name, the more valuable it will be. It should describe the technology and be easy for people to remember and type in a browser. If I had a time machine, I'd go back to the beginning of the Internet age and buy up all the simple domain names that are worth big money now, e.g., www.car.com, www.food.com, and www.sex.com.

Corporations and Trademarks

Another inexpensive way to begin to protect your IP is to form a corporation. Forming a corporation usually costs only about $50 to $100 in most states and you can do it via the Internet. Choose a name that describes your technology and that can eventually become your trademark once it is used in interstate sale of the invented product. Once you have a company name, use a phrase describing your technology ("branding") and mark it with the "TM" trademark symbol, as long as no one else is already using it—also something you can search through the respective state business registration sites and the U.S. Patent and Trademark Office.

Particularly effective are corporation names that describe the product, like Plax, the antiplaque mouthwash, or Pet Rock. For added protection, you can register your trademark nationally. In many cases, the trademark may ultimately be worth more than a patent since it lasts forever and a patent only lasts twenty years! Trademarks usually cost about $1,000 or more through a patent attorney, depending on how many categories you file for, e.g., the product itself, T-shirts, movies, and toys. Each category has a fee.

Copyright

Copyright is another form of protection; however, it is designed for orginators in the realms of print, performance, film, music, etc. For example, we, as authors of this book, own the copyright to our creation. For added protection we may also trademark the name and, of course, obtain a domain name for the Web site (www.hardcoreinventing.com).

For most of my inventions, I try to establish as much IP as I can afford. I usually start with the domain name, start a company, register the trademark, and get a patent. In the case of PocketFloat technology, I obtained the domain (PocketFloat.com), formed a company (PocketFloat Corporation), and obtained a patent (Emergency Supplemental Floatation Device, U.S. Patent #7,104,858). I now have all the IP I need to make it a success, either by selling the product, selling the company, or licensing the IP.

It is important to note that IP is useful in two distinct ways: First, it protects you from people copying your invention, essentially giving you a monopoly on your technology if you obtain broad patent coverage; second, it also represents equity. IP represents value and can be sold, so you want to build up your IP to possibly sell or license. With this in mind, it makes sense to develop IP for protection and marketing purposes. This is all good.

SUMMARY

- Build up your IP for protection of your invention and marketing purposes.
- Consider applying for a utility patent.
- Get a domain name.
- Form a corporation.
- Trademark your product.

Trademark Infringement Claim—Scrabulous

Hasbro, the world's second-largest toy and game company, owns the rights to the crossword board game Scrabble in the U.S. and Canada, and Mattel Inc. owns the rights elsewhere. In 2008, both companies demanded Facebook remove Scrabulous, an online version of the game, and jointly issued cease and desist notices to four parties involved in the development, hosting, and marketing of Scrabulous.

The Scrabulous application, developed by brothers Jayant and Rajat Agarwalla of Calcutta, India, reportedly had 600,000 Facebook users daily. Lawyers say that Hasbro appears to have a valid trademark infringement claim against the Scrabulous developers because the word is close enough to Scrabble to confuse people. Jessie Beeber, a New York attorney, declared, "There's a psychic harm to consumers because they're being lied to and misled into thinking there's some link here."

Legal opinion is that a copyright claim might be harder to prove because it's not clear that board games like Scrabble can be copyrighted, but that courts tend to be sympathetic if it seems like game makers are being abused. However, even if the Scrabulous developers have violated a copyright or trademark claim, the application won't necessarily disappear from Facebook soon. Proving a case or enforcing a judgment against the Agarwalla brothers in India could be difficult given the differences in intellectual property law between India and the United States. Also, it's not certain that Facebook would be liable for contributing to copyright or trademark infringement.

For Facebook, however, this action by Hasbro could be the beginning of legal skirmishes if other applications on the site infringe on companies' intellectual property rights. As a result of the legal matter, Scrabulous fans can find their favorite game on www.Lexulous.com.

Copyright Infringement Claim—Fatso the Ghost and the Ghostbusters Logo

In 1984, Harvey Productions, the home of *Casper the Friendly Ghost*, sued Columbia Pictures over the movie *Ghostbusters*. They complained that the cartoon ghost in the logo of Bill Murray's crew looked a lot like Casper's friend, Fatso. Columbia Pictures convinced a judge that a lapsed copyright had put Fatso into the public domain and this ended the case. Then researchers discovered that Harvey Productions had failed to renew other copyrights covering the company's ghosts. It seemed Casper could be public property, too.

Copyright Infringement Claim: Barbie vs. Bratz

Barbie may have taken some of the pout out of her bratty rivals Yasmin, Jade, Cloe, and Sasha: A federal jury awarded toy giant Mattel Inc. $100 million in damages in a federal copyright lawsuit against Bratz maker MGA Entertainment. And in November 2008, attorneys for Mattel Inc. asked the court to ban MGA from making the dolls and to impound and destroy the Bratz product. Earlier, the federal jury ruled that the Bratz doll designer had come up with the edgy concept while working for Mattel Inc.

It is still a matter of dispute whether only the first generation of Bratz infringes Mattel's copyrights or whether all the dolls in the line are in violation.

Step-by-Step Guide From Disclosure to Patent— Etch It in Stone

Congress shall have the power to promote the progress of science and useful arts, by securing for limited times to authors and inventors the exclusive right to their respective writings and discoveries.

—Article 1, Section 8, U.S. Constitution

ALL ABOUT PATENTS

PATENTS WERE INTENDED TO CREATE A REPOSITORY OF how-to information. Since Thomas Jefferson handed out the first patent in 1790, the U.S. Patent and Trademark Office has granted more than 7,450,000 patents in an effort to encourage scientific advancement and economic prosperity. Society benefits from well-made, constantly improving products. When a patent is granted, the invention becomes the property of the inventor and, like any other form of property or business asset, can be bought, sold, rented, or hired. At the patent office you can study existing technology. Many inventions are improvements on existing technology as an invention is a constant work-in-progress.

US005421287A

United States Patent [19]

Yonover

| [11] | Patent Number: | 5,421,287 |
| [45] | Date of Patent: | Jun. 6, 1995 |

[54] **VISUAL LOCATING DEVICE FOR PERSONS LOST AT SEA OR THE LIKE**

[76] Inventor: **Robert N. Yonover**, 219 Koko Isle Cir., Honolulu, Hi. 96825

[21] Appl. No.: **152,349**

[22] Filed: **Nov. 17, 1993**

[51] Int. Cl.⁶ **B63B 45/00**
[52] U.S. Cl. **116/209;** 441/83; 116/26
[58] Field of Search 116/209, 210, 211, 26; 441/11, 36, 83, 89

[56] **References Cited**

U.S. PATENT DOCUMENTS

2,842,090	7/1958	Samwald	116/210
2,949,094	8/1960	Clothier	116/209
3,002,490	10/1961	Murray	116/210
3,877,096	4/1975	Scesney	116/63 P
3,952,694	4/1976	McDonald	116/209
4,418,733	12/1983	Kallman	383/11
4,809,638	3/1989	Kolesar et al.	116/26

FOREIGN PATENT DOCUMENTS

| 468051 | 1/1975 | Australia | 116/26 |

| 5319375 | 12/1993 | Japan | 441/89 |

Primary Examiner—William A. Cuchlinski, Jr.
Assistant Examiner—Willie Morris Worth
Attorney, Agent, or Firm—Shlesinger Arkwright & Garvey

[57]   **ABSTRACT**

A signalling device for indicating, by day or night, the position of a person lost at sea (on land or in space) comprises an elongate brilliantly colored streamer made up of flat, flexible, inherently buoyant material with built-in support struts to keep the material at maximum outstretched surface area. The streamer can be coated with any one or more of the following in any combination: brilliant color, phosphorescent pigment, reflective material, or International Distress Signal indicia. The device may be attached to a life jacket and rolls up into a water-release container secured to the life jacket. Upon deployment, the container converts into a sun-protective, radar-visual reflective, and water catchment hat. The streamer is extended manually or automatically and can remain in an outstretched manner indefinitely.

26 Claims, 2 Drawing Sheets

Types of Patents

There are three types of patents:

- Utility patents cover those things most people usually think are covered by patents, a new mousetrap, a new process, etc. Utility patents cover four types of inventions:

  1. Machines: Mechanical devices made of parts that work together to do something. Machines can be new or composed with existing inventions to achieve a new purpose.

  2. Compositions of matter: Chemical compositions, mixtures of ingredients, as well as new chemical compounds. These include drugs, chemicals, or metallic alloys.

  3. Processes: A mode of treatment of certain materials to produce a given result—a new way of doing something.

  4. Man-made products: Anything you can make that isn't a machine or a composition of matter.

- Design patents cover designs in manufacturing or building, based on the unique design and appearance of the item.
- Plant patents protect certain types of plants.

Things You Can't Patent

- You cannot patent an idea or suggestion. You need a complete description of the actual machine or other subject matter for which the patent is sought.
- Patent protection is not available for natural processes you discover, such as the law of gravity, or mathematical principles. You can patent something that uses natural or mathematical laws (e.g., computer software) but you can't try to stop other people from using those laws in the future.
- You can't patent a natural product, such as a rare medicinal herb.
- A utility patent cannot protect printed material, even though it is a man-made item.
- Some printed matter can be protected by a design patent, but only if the design of the material, and not the written content, is being patented.

What Can Be Patented?

The patent law specifies the general field of subject matter that can be patented and the conditions under which a patent may be obtained. Any person who invents or discovers any new and useful process, machine, manufacture, or composition of matter, or any new and useful improvement thereof, may obtain a patent subject to the conditions and requirements of the law.

For a patent application to succeed, your invention must be novel (it must not have been made public in any way), inventive and nonobvious (the invention must not be obvious to someone with a good knowledge of the topic area), and have industrial utility, i.e., the invention must be useful in providing some level of benefit. Inventions are things and need detailed descriptive drawings that show how they work. (Software is a special case, and different rules apply.)

Novelty Checklist

Your invention must be new and different from anything that has been available to the public before. It cannot be patented if the invention was known or used by others in this country before the invention by the applicant for the patent. The invention is also not novel and cannot be patented if it was described in a printed publication in this or a foreign country, or in public use or for sale in this country more than one year prior to the application for patent in the United States.

If the inventor describes the invention in a printed publication or uses the invention publicly, or places it on sale, he must apply for a patent before one year has elapsed, otherwise any right to a patent will be lost. The inventor must file on the date of public use or disclosure, however, in order to preserve patent rights in many foreign countries. You can find out more about the novelty requirement at the Web site of the U.S. Patent and Trademark Office (www.uspto.gov). And be sure to note their words of warning: "Inventors are reminded that any public use or sale in the United States or publication of the invention anywhere in the world more than one year prior to the filing of a patent application on the invention will prohibit the granting of a U.S. patent on it. Foreign patent laws in this regard may be much more restrictive than U.S. laws."

Nonobviousness Checklist

Even if the subject matter sought to be patented is not exactly shown by prior art, and involves one or more differences from the most nearly similar thing already known, a patent may still be refused. The subject matter sought to be patented must be sufficiently different from what has been used or described before, such that it may be nonobvious to a person having ordinary skill in the area of technology related to the invention. An invention that solves a known problem is nonobvious if others failed in their attempts to solve the problem. The patent office will ultimately decide if the invention is obvious or nonobvious. See the USPTO's Web site for more information.

Utility Checklist

Your invention has to be useful for something. You can't get a patent for a process that produces something for which there is literally no use, or for a drug that has not been fully tested, or for an invention whose sole purpose is immoral or illegal.

Confidentiality and Publication

To obtain a patent the invention must be novel, and if the inventor makes it public at any time before submitting an application, then it is clearly no longer novel. This can cause the patent application to fail. Take care what you write in e-mails or put on your Web site, and keep all details out of conversation with others.

For an academic, publishing papers that describe a new technology or process before a patent application is filed will almost certainly destroy the chance of a patent being granted. Once you have filed the application for a patent, you can publish safely in the United States. However, it is always wise to check first with a patent attorney, particularly if you will be seeking international patents for your process.

Describing the invention to others may also count as a public disclosure, so it is very important that the details of the invention are not disclosed or that there is a confidentiality agreement in place to protect your product idea. A confidentiality agreement will not only bind the other party to secrecy, ensuring that the disclosure does not prevent patenting, but will also restrict the other person from using your invention. It is a simple statement to be signed by anyone seeing the details of your invention. It states that they have seen your invention and agree not to tell anyone else about it.

As the patent goes to the person who made the discovery first, make sure you keep meticulous records of the steps taken to arrive at the invention. Again, make sure you keep an inventor's log. Use detail. Get it notarized. Use pictures. When you're ready to proceed on getting a patent, send a copy of your notes to your patent agent or attorney.

If you do finally file a formal patent, the examiners will perform their own search as part of the patent application filing fees; however, in some cases it is worth performing a formal search or paying for a professional search by an agent. This could save you thousands of dollars if it saves you from filing a formal patent application when the invention or a conflicting patent already exists.

How to Do a Preliminary Patent Search

Before you decide to move ahead with patents, it is quite easy to do a preliminary patent search to see if your product idea is unique or if someone has already patented it. Patent attorneys do professional searches (and opinions) for a price; however, it is much cheaper to at least first carry out a search yourself.

Searches can be done on the Internet through the U.S. Patent and Trademark Office. Google also has an excellent patent search engine at www.google.com/patents. Another good site is www.freepatentsonline .com, which shows copies of PDF documents containing patent images.

The art form of searching is based on using key phrases that describe your invention, e.g., a RescueStreamer technology search might start with the key words, "streamer," "visual," "target," etc. Be patient. Searching for similar patents is a time-consuming endeavor as sometimes the list will contain hundreds of patents if the term or phrase you have selected is popular. Once you have a list of similar-sounding products, you need to read the patents and see which ones seem relevant. You can improve search quality by using the USPTO advanced search page and searching in the specification field. If you can't find anything close to your invention, that's good news—but that doesn't mean it hasn't been done already. If you are not familiar with advanced searching techniques, you are quite likely to miss what you are looking for. But keep in mind, a professional search by a patent agent costs around $500 or more.

What if You Find an Existing Patent?

If you do find an existing patent, after the initial depression celebrate the fact that you saved money by not pursuing a patent and that you did come up with an excellent product idea—it's just that someone else thought of it first.

One further option is to try to contact the patent holder and see if anyone is doing anything with the patent. Statistically, most people do not end up building or profiting from their invention. Check if the inventor is interested in working together on pushing the technology ahead to the marketplace.

Who May Apply for a Patent?

According to the law, only the inventor may apply for a patent, with certain exceptions. If a person who is not the inventor should apply for a patent, the patent, if it were obtained, would be invalid. The person applying in such a case who falsely states that he is the inventor would be subject to criminal penalties. If the inventor is dead, his legal representatives, i.e., the administrator or executor of his estate, may make the application. If the inventor is mentally challenge a guardian may make the application for patent.

If an inventor refuses to apply for a patent or cannot be found, a joint inventor or, if there is no joint inventor available, a person having a proprietary interest in the invention, may apply on behalf of the nonsigning inventor. If two or more persons make an invention jointly, they apply for a patent as joint inventors. A person who makes only a financial contribution is not a joint inventor and cannot be joined in the application as an inventor. It is possible to correct an innocent mistake such as erroneously omitting an inventor or erroneously naming a person as an inventor.

Should You Patent it Yourself?

Once you have determined that your product idea or invention is worth protecting, it is time to get serious on the legal front. The inexpensive yet usually weak (you-get-what-you-pay-for-method) is to try to patent it yourself. An excellent resource for this is *Patent It Yourself* by David Pressman. I actually tried this for my RescueStreamer technology after reading the Pressman book. Of course, it was rejected, as most are at first, at least most of the "legal claims." The technology was not perfected yet anyway so it was not the right time. It was an excellent exercise and helped me become skilled in writing the text of a patent application.

Regular Patent Application for a Utility Patent

The two main parts to the regular patent application are the specification and the claims.

1. The specification: This is a written description of how to make and use the invention in sufficient detail that a reasonably skilled person in the field could reproduce it from the description. If necessary, drawings or diagrams can be attached. These specifications must be absolutely precise.
2. The claims: These are the aspects of the invention that you claim are new and inventive. Once you get a patent, the scope of coverage will be defined by the wording in the claims, so you must be extremely careful and precise when you write the claims. In the claims, you should include everything that you think is unique to your invention and, more importantly, what you want to have rights to make, use, and sell such that you can exclude others from doing so. This is the most difficult part of filing a patent. It is very important to take great legal care so that you adequately protect yourself and your invention. Typically you need a competent patent attorney to get the best patent coverage for your invention.

You must be totally honest in your application. The USPTO imposes a "duty of candor" on all applicants, which means that if they determine that you failed to provide relevant information when you were applying, they can deny your patent on that ground. Or, perhaps even worse, deny you patent protection in later litigation.

The USPTO's Web site provides information and some forms. They provide detailed step-by-step explanations of how to complete and submit the applications. As you read their Web site, you will note that applying for a patent is a complex and involved task. The patent office recommends using an experienced attorney or patent agent.

If your patent is accepted, you can spend a while congratulating yourself. If not, you will be told how to go through the appeals process when you receive the rejection.

Using a Patent Attorney

Although now I go the formal route and pay the big money to the patent attorneys ($5,000 minimum), I usually write the patent application myself, which saves some money. However, I have the patent attorney write the patent claims. Claims are the legalese that represents the meat of your legal protection. It is not worth cutting corners on the claims—work out a deal if you are short on cash—but make sure the claims for your invention are written by a patent attorney. Pick a patent lawyer technically trained in your field of invention and knowledgeable about the marketplace for your particular product. The *Martindale-Hubbell Law Directory* can be found in many libraries and will tell you the areas of expertise for patent lawyers listed.

Patent Tips from Neustel Law Offices Ltd.

A very important part of the patent application are the "claims," which describe the scope of coverage the inventor is attempting to receive from the U.S. government. A competent patent attorney is typically required to receive the most favorable patent coverage for your invention. Adequate patent coverage ensures that potential infringers will be prevented from making, using, or selling your invention even if they make a modification.

After filing the patent application with the USPTO, an office action from the USPTO will usually be received within eight to fourteen months. Typically, the USPTO will reject some or all of the claims of the patent application depending upon whether the USPTO examiner believes it would have been obvious to create your invention because of the prior art located by the examiner.

It is then necessary to argue that your invention is patentable based upon the differences between the invention and the art cited by the USPTO examiner. Legal arguments and decided case law may be used to refute the examiner's position. A telephone interview with the examiner may also be arranged to find agreement on any issues of dispute.

Virtual TV or VTV

Two young computer enthusiasts, Susan Wyshynski and Eric Gullichsen, were living on a houseboat when they invented Virtual TV or "VTV," the technology that lets you pan around a 360-degree photo or video. Without any great source of funds, all research and other expenses came out of pocket.

They knew they had a good idea and decided to go all out and patent it, so the young inventors chose the best patent attorney they could find. He specialized in the relevant field and he'd acted as patent attorney for Apple. But he charged $450 an hour. They walked into the lavish legal offices, and the attorney began chatting.

Eric interrupted immediately. "Excuse me," he said. "Are you charging us for this? Because once we're on the clock, we're only talking about what we're paying you to talk about, our patent process and that's it. Any time you're charging us, we don't want to have any conversations!"

The initial paperwork took six months. Eric and Susan did their own drawings to save money. When the invention was in the filing process, drafts went back and forth from the inventors to the attorney as he added amendments to cover all aspects of the process. The rewriting was taking a long time, but Eric and Susan thought they were ahead of the game.

Then, visiting a Las Vegas computer expo, they walked into the megasized Microsoft booth and were horrified when they saw their invention projected on a huge movie-size screen and advertised as a coming feature to Microsoft technology. Susan says, "We thought we were screwed. We had no way of finding out what Microsoft's status was." They immediately contacted the attorney, telling him they needed to move fast. They started praying that they could get to the "patent pending" stage before their competitors did.

The patent took four years and cost $20,000. Expensive, but worth it as the competition in the field was huge with Apple, Microsoft, and five other big-name companies all busy developing the technology. Susan and Eric's patent was the only one that could hold up in court. The attorneys of the major players couldn't get around it. The technology was sold to Be Here Now and the inventors left to do other inventive things—like jetting around the world checking out surf breaks.

Titan Key Software

Titan Key Software, the brainchild of inventor and entrepreneur Peter Kay, is the first technology that stops spam at its source. Most other spam blockers accept the evil e-mails and place them in quarantine or some other folder. Users still have to dump the e-mail or deal with it later. Titan Key never accepts the spam in the first place.

Kay says getting patents for his invention was no easy task. Ensuring that the unique features of his spam filter invention were sufficiently protected both in the United States and internationally took six years. During that time, Kay says, he spent dark nights questioning whether it was worth risking his family's money and even questioning his self-worth. Happily—it really was!

Titan Key had a patent pending from 1999. During that time, Kay proceeded with plans to license Titan Key's software himself and expand the company. This strategy, he says, requires little venture capital and is ideal for small startups. He established a customer base globally, and then one of the international customers, a prominent businessman, fell in love with the product and introduced Kay to a large Singapore-based technology group. Titan Key Software was purchased in 2007 for more than $2 million. But then that deal—a complex deal involving a third party—fell through.

Kay persevered. He found a broker good at lining up intellectual property deals. The broker found a company that buys patents and closed a deal in 2007.

Kay believes in giving back. Entrepreneurial firms, he says, represent 10 percent of new businesses in the United States, and yet account for 75 percent of new noncorporate jobs. With the desire to help other entrepreneurs, he formed Flat Earth Ventures (www.flatearthventures.com) as a high-tech mentoring company. His advice to inventors is to stick with it, persevere, and try every angle.

Alternatives to Utility Patents

WHAT IF YOU ARE NOT SURE YOUR INVENTION WARRANTS the cost of a patent? Many attorneys and patent agents will give inventors a free consultation to determine the cost of applying for a patent. Call to check first. The patent process can take a long time and cost thousands of dollars. Also, if an inventor keeps improving on his original idea, he ends up with a patent only on the original and not on the new improved version. Discuss this also with the patent attorney or agent.

According to Stephen Key, an inventor and cofounder of InventRight, a company that sells educational CDs to inventors, most inventors spend too much time and money trying to protect their ideas, and not enough selling them. Key advises against rushing into the lengthy and costly patent process, because 97 percent of the patents he sees never make any money. Instead, he says, inventors can protect their ideas as "patent pending" with a provisional patent for a $110 fee, and tweak the product while marketing it to manufacturers.

"It gives you one year to kind of fish off the pier," says Andrew Krauss, Key's business partner and president of the Inventors Alliance, a California-based educational and networking group for inventors. (See more at www.inventright.com.)

PROVISIONAL PATENT APPLICATIONS

Since June 8, 1995, the U.S. Patent and Trademark Office has offered inventors the option of filing a provisional patent application. This is designed to provide a lower-cost first patent filing in the United States and to give inventors time to develop and test a product idea, disclose it safely to vendors and others, and determine whether the invention merits further investment in a twenty-year patent. Call 1-800-786-9199 for a free brochure from the U.S. Patent and Trademark Office.

Advantages of Filing a Provisional Patent Application

One of the main advantages of filing a provisional patent application is the lower initial cost of preparation and filing. Other advantages are:

- It provides the means to establish an early effective filing date in a patent application and allows the term "patent pending" to be applied. This gives you significant marketing advantages and twelve months to market your invention and decide whether you should move forward with the more expensive regular patent application.
- It establishes an official U.S. patent application filing date for the invention and permits one year's authorization to use a "patent pending" notice.
- It enables immediate commercial promotion of the invention with greater security against having the invention stolen.
- It allows for the filing of multiple provisional applications for patent and for consolidating them in a single, nonprovisional twenty-year patent application.

The provisional patent application can be filed up to one year following the date of first sale, offer for sale, public use, or publication of the invention. Note: These pre-filing disclosures, although protected in the United States, may preclude patenting in some foreign countries. You should consult with a patent attorney if you have any questions.

Although drafting requirements for the provisional patent application were originally fairly forgiving, recent court decisions indicate that any laxness in drafting a provisional patent application can be fatal to rights granted later. To be certain your provisional patent adequately protects your invention, have it prepared by a patent attorney. This is a tricky field and you do not want to make a costly mistake. Also advise the patent attorney if you are contemplating applying for patents in foreign countries.

A provisional patent application should contain:

1. A written description in full, clear, concise, and exact terms covering the full scope of the invention and the manner and process of making and using the invention. Note that when you file your regular patent application (within a year after filing for the provisional patent) your description must not contain any new technical information that was not in the description submitted with the provisional patent. You cannot change the product midstream. So you have to know exactly what you're doing before you file either application.
2. Any drawings necessary to understand the invention. Drawings must support the subject matter of your invention in order to later establish priority. Drawings should be technical drafter's drawings with exact specifications and measurements.
3. The names of all inventors.
4. Filing fee and a cover sheet identifying the application as a provisional patent application. The cover sheet provides all of the information needed to process the provisional application promptly and properly and prepare the filing receipt.

The written description and any drawings for the provisional application do not have to be identical to, but must adequately support, the subject matter claimed in the later-filed regular patent application in order to benefit from the provisional application filing date.

A provisional patent application will remain pending at the patent office for twelve months from the date it is filed. This twelve-month period cannot be extended. An applicant who files a provisional application must, therefore, file a regular patent application during the twelve-month period in order to claim benefit of the earlier provisional patent application filing date.

Missing the deadline can be catastrophic—an applicant whose invention is "in use" or "on sale" in the United States during the twelve-month patent pending period may lose the right to patent the invention if that twelve-month period expires before a regular patent application is filed.

The USPTO Web site gives specific details and provisional patent forms that you can download.

FOREIGN-COUNTRY PATENTS

Without a patent in a foreign country, you cannot prevent your product from being copied overseas. Your U.S. patent, however, can help you (with the assistance of your patent attorney and U.S. Customs) to prevent foreign-made copies from being imported into the United States. To stop your product from being copied in foreign countries, you will need foreign-country patents.

The European Patent Convention (EPC) is made up of thirteen countries and the EPC Office performs an initial search and examination on your submitted patent. Your patent must then be submitted to the specific countries you have chosen with the necessary fees. Each European country will issue you its own national patent.

For patent coverage in countries outside of Europe, you may use the Patent Cooperation Treaty (PCT). Within twelve months after you file your U.S. patent application, you file the same application for a PCT international patent at the USPTO Receiving Office, indicating in which of the seventy-four member countries you wish to obtain patent coverage. You can file using the PCT within the twelve months to obtain your effective filing date, and wait until later to decide in which countries you want to proceed. Or you can file directly in those countries. Your patent attorney or agent will be able to give you an estimate of costs for foreign patents.

When filing for foreign patents, remember another difference—the United States is a "first to invent" country, but all other industrial countries are "first to file."

Patent Tip from Amanda Budde-Sung, PhD, Expert in IP

Amanda Budde-Sung, PhD, expert in international management, intellectual property, and biotechnology, advises that a U.S. patent gives the greatest amount of protection with fewer morality hoops to jump through—one of the reasons the United States is a leader in biotechnology.

An example is the Harvard OncoMouse, a genetically engineered mouse. Patent applications on the OncoMouse were filed in the mid-1980s. The United States granted a patent, but Europe and Japan at first refused patent coverage because the mouse is a man-made life-form. In Canada, the Supreme Court stopped the mouse in its tracks for more than a decade, opposing the patent on the grounds of morality.

ALTERNATIVE WAY TO OBTAIN A PATENT

Ed Zimmer, successful inventor and entrepreneur, is the president of the Zimmer Foundation, a public service organization counseling inventors and entrepreneurs on product licensing and starting a business.

Zimmer says that inexperienced inventors generally do not have a realistic handle on the marketability of their invention—but their licensing prospects do. On the following pages is Zimmer's "business bluff" approach, for inventors wishing to obtain a licensee. It offers an evaluation of the invention (by the ultimate decision makers) before going on to spend the money on patents that may never be licensed. It provides a rational approach for inexperienced inventors hoping to license their inventions who are unable or unwilling to invest in the services of a registered patent attorney or agent.

TIP

Tip from Ed Zimmer, The Zimmer Foundation

Ed Zimmer's Business Bluff Approach:

If you're thinking that companies might go looking through issued patents to find new products—they don't. The only reason they look through issued patents is to find "prior art" (i.e., already patented features and methods) that they must design around in the development of their own new products.

Does that mean that patents are a waste of time and money? Generally, "yes." What good will a patent do you if your product proves to be not licensable? A patent will cost you several thousand dollars—and your risk/reward ratio is in the same neighborhood as "investing" it in your state lottery. (The lottery's odds are a bit longer, but its jackpot is greater.) And don't expect a patent attorney to tell you whether you can get a sufficiently broad patent—only you can determine that. The attorney simply doesn't have the market knowledge to assess what's important in your market.

So we're back to where we were before the provisional came into law: spend a lot of money to safely show a new product to industry—that they probably don't want. There is a solution.

In the "old days" (before the provisional), cost-conscious inventors would try to arrange with their patent attorney to hold their important intellectual-property papers in the attorney's file, and then contact their licensing prospects to see if there was any interest. This was a relatively "safe" method because if a company was interested, and knew that a patent attorney was involved, it would be extremely unlikely that the company would try to do anything with the product idea without at least talking with the patent attorney. And if it appeared that useful patent protection could be obtained, the attorney and the company would work together to ensure that the best possible protection was obtained (and that the inventor received a "fair" deal). This is still a fine approach to licensing.

If you're prepared to research your product idea fully and thoroughly, a patent (or more likely patents) may make sense. But do the research first, as your research will most likely show you why your product idea isn't licensable. And even if it appears to be licensable, you'll find that the time and costs to make it licensable will be substantial—much greater than simply the patent costs.

When the provisional application came into being, I had hopes that people with new product ideas might finally have a way to safely show their ideas to industry at a reasonable cost. But that has not proved to be the case. For a provisional to offer any "protection," it must be both enabling (i.e., must completely and accurately describe how to make and use the device) and thorough (i.e., it must also describe the device broadly enough to provide adequate support for all of the patentable features of the device in a later nonprovisional application).

Evidence indicates that most people trying to use the provisional cannot do this even close to adequately—and if they go to a patent attorney to draft it for them, they find that the cost is almost as much as filing a nonprovisional patent application (and many practitioners will, rightly, urge them to bypass the provisional and simply file the nonprovisional).

However, you'll find it difficult to find patent attorneys who will work with you in this way—you're a "nuisance" to their normal workflow. And even if you do find one, they will likely insist that you first pay them to do a professional patent search. If one of the companies you contact does contact the patent attorney, he or she needs to be prepared to discuss with the company exactly what might be patentable. And you may find it awkward, when the licensing prospects ask whether your product idea is patented, to explain that, no, it isn't, but that the important intellectual-property papers are in your patent attorney's files.

Although the provisional does not provide the low-cost vehicle that I had hoped it would, its existence in current law does open up a variation on the old approach that avoids the difficulties. Here are the steps:

The "Business Bluff" Approach

1. Compile a list of your licensing prospects.
2. Write up and file a provisional application. It doesn't have to be a "good" write-up because you're never going to show it to anyone.
3. Contact all the companies on your "prospects" list. When they ask whether your product idea is patented, say that it's "patent pending." If they ask if that's by provisional or nonprovisional, say "provisional." If they ask what the product idea is, tell them as much as they want to know—there's nothing to gain by holding anything back (and much to lose if you do). If they want to see more information, send them a clear and complete description of the product or improvement—but not one that looks like it might have been written as a provisional application.

If none of your prospects are interested, it's only cost you the provisional application filing fee (currently not more than $100) and a couple of months of e-mails and phone calls. Hopefully, one of the prospects will be interested. They'll ask to see your provisional. Do not show it to them—put them off by saying something like, "I've made some design changes since the provisional, so I won't be citing the provisional in my formal application." (Never give any indication that you might not be following through with a formal application.)

Then negotiate with them to pay for the formal application. The cost to them is trivial compared with the other costs they face in bringing the product to market—and if a broad patent is possible, that's as much in their interest as in yours. (The patent will be in your name, as only the true inventor can file for patent.)

If the company likes your product and wants to bring it to market, you've put them in a bind:

- They can't go ahead without you—because they don't know what you're patenting—they have to assume it's the broadest patent possible, and they have to assume that you'll be following through with your formal application, so that at about the time they get the product to market and it's starting to make money for them, they have to expect that they'll be faced with your patent issuing.

- And they can't "wait you out"—because a patent can take several years to issue—and it will be a long time (probably past the market window) before they can be sure you didn't follow through with your formal application. Their only rational business decision is to work with you.

Understand that this approach has nothing to do with intellectual-property law (other than that the existence of the provisional in the law makes it possible). What you're doing here is running a business bluff, but a bluff in which no one can ever see your hand unless you accidentally show it or you allow someone to trick or coerce you into showing it.

The people you'll be contacting are pros—they're fully capable of bluffing back. You're very likely to hear, "We're very interested in licensing your idea—but we can't move forward until we see your provisional." You have to be prepared to simply say good-bye. They weren't really interested in licensing at all—all they wanted was to see what you were planning to patent.

To summarize: Given a product that you're not prepared to "venture," you really have only three choices that are at all rational:

1. Go about it "right." Do the research and prepare a presentation that offers not just an invention but also a documented, substantiated profit-making opportunity. This is the approach used by the pros—the "professional" inventors. It provides by far the greatest odds of successful licensing—not because the pros are necessarily more creative, but because as they research their ideas and encounter the obstacles that are always present, they improve their inventions to overcome or bypass those obstacles. Hence, when they're ready to present, they have something they know will license, and generally to whom and why.

2. Use the "business bluff" approach outlined above. The odds of successfully licensing are obviously much lower than the first choice (but certainly better than not trying to license at all). This isn't likely to get you a license with any of the giant companies (neither

is any approach other than number one, above.) But it may well get you a license with a company that is not the market giant—and it may get you a license even if there are serious patent obstacles. If you present them with a patent (or provisional) that they can design around, they will very likely do so. But if you leave the patent issue up to them, they may well decide that they don't care about patent protection—that first-to-market is enough for them. And even if they do care, you've given them the opportunity to get the broadest possible patent protection—which it's unlikely you would have discovered (or have paid for) on your own.

3. Forget about the whole thing and get on with your life.

Any approach you try between the extremes of choices one and two will almost certainly be a waste of time and money. Yes, you may well get a license with only a half-baked patent and a mere submittal of your "invention"—but if that's the case, you would have obtained the same license with choice two, without the front-end costs and hassle.

Note: This advice refers to inventions and not ideas. Ideas are not licensable, only inventions are. Do not use choice two to try and license an idea. (See more at www.tenonline.org, The Entrepreneur Network.)

WHAT IF SOMEONE USES MY PATENTED INVENTION?

Even if your invention is patented, you may still come across someone who is using your product idea without getting your permission, either on purpose or by accident. It is up to you to enforce your rights. If you suspect someone is making, using, or selling something you've patented without paying you a licensing fee, you must tell him or her to stop immediately. If this has no effect, it is time for you to consult an attorney who will have a heart-to-heart talk with the foul infringer, and also possibly the company buying products from the foul infringer, advising that you intend to enforce your patent rights to the fullest.

You may have had an idea for a product, drawn diagrams, had it patented, but not had the money to actually build the product. In this case, your patent is still good and if someone else makes the exact same product, you're still protected. The patent infringer should be talked into paying a licensing fee or to stop infringing your patent. If the infringers won't settle, you should talk to an attorney about taking them to court.

TIP

Tip from David Pressman, Patent Attorney and Author of *Patent it Yourself*

In his book, *Patent it Yourself*, David Pressman warns never to assign or transfer ownership of your patent in return for a series of payments: If your assignee defaults on the payments, you'll be left without your patent or your money, but with a big legal headache—getting your patent back. If someone wants to buy your patent, see a lawyer or use a legal forms book to make a suitable license with an agreement to assign only after all payments have been made.

Patent Infringement—Guitar Hero

The enormously popular video game Guitar Hero has become the subject of some interesting and expensive patent-infringement lawsuits. In a letter from their attorneys on January 7, 2008, iconic guitar manufacturer Gibson Guitar Corporation accused video game giant Activision of taking advantage of Gibson's patented technology without properly compensating Gibson. They requested that Activision obtain a license under the patent or halt sales of any version of the Guitar Hero game software. Gibson stated it owned and will defend a nine-year-old patent for "technology to simulate a musical performance."

On March 20, 2008, Gibson also filed patent-infringement lawsuits against Harmonix, MTV Networks, and Electronic Arts. The court will determine the outcome of these lawsuits. Another lawsuit, this one for millions of dollars in claimed unpaid royalties, was brought by Harmonix against Activision and settled out of court.

Jerome Lemelson

An inventor who tenaciously pursued his rights to his patented products and innovations, Jerome Lemelson was rewarded for his patents and his technological foresight to the tune of more than $1.5 billion in licensing royalties.

One of America's most prolific inventors, Lemelson held more than 600 patents, including claims for products from camcorders, fax machines, cassette players, bar-code scanners, and automatic teller machines to crying dolls. Lemelson's inventions were often so far ahead of their time that the technology required to build them did not yet exist. Nonetheless he filed patent applications that remained pending for decades and the delay worked to his advantage.

His patent-procurement strategies were criticized by the companies he sued for infringement as creating "submarine patents," because many applications were pending for years. In fact, whole industries by some of the world's largest companies were built around a technology later alleged to be covered by one or more of his patents. And the companies paid dearly.

In 2005, the Federal Circuit Court ruled that Lemelson's eighteen- to thirty-nine-year delays in prosecuting patent claims relating to machine vision and bar-code technologies were unreasonable. With machine vision, Lemelson had submitted his original application in the 1950s. His first patent on the technology was issued in 1963, and he filed continuation applications and was granted additional patents in the 1980s when the technology was in widespread use.

Lemelson, who died in 1997, defended the rights of the individual inventor, stating that "company managers know the odds of an inventor being able to afford the costly litigation are less than one in ten; and even if the suit is brought, four times out of five the courts will hold the patent invalid. When the royalties are expected to exceed the legal expense, it makes good business sense to attack the patent. . . . We don't recognize that the consequence of the legal destruction of patents is a decline in innovation." Criticized by some and applauded by others, this incredibly successful inventor and innovative thinker for decades successfully defended his patented innovations. Lemelson and his heirs also encouraged and continue to help new inventors by donating $130 million through the Lemelson Foundation, the Lemelson MIT Foundation Program, and other foundations.

WEIRD AND WONDERFUL PATENTS

As a break while pondering patents, you can find all sorts of weird, wonderful, and inspirational patents online at www.freepatentsonline.com. This site also gives detailed descriptions of millions of inventions.

Patent 395,515:1889 Gum Locket. A locket, invented by C. W. Robertson of Tennessee, is worn as a piece of jewelry on a chain that's padded inside and fitted with a concave piece of crystal serving as an anticorrosive lining. Chewing gum "is not left around carelessly to become dirty or to fall into the hands of persons to whom it does not belong."

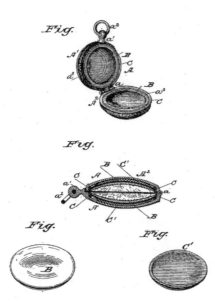

SUMMARY

Reasons for Patent Protection

- Protects patent owner from competitors copying the patented innovation or invention
- Protects research and development expenditures
- Protects from misappropriation by patent owner's employees and others.
- Legally hampers and discourages competition.
- Can be used as a bargaining chip in settlement of controversies.
- Discourages competition.
- The invention is in a field in which competitors traditionally obtain patents.
- Increases patent-owner company's worth
- The patent is an asset that will create incoming licensing royalty income.
- Documents innovation to patent owner's customers and contributes to the sale of products and services
- Public disclosure of the invention may soon be required.
- The invention is likely to be independently invented by others in the future.

Reasons Against Patent Protection

- Product may be of limited commercial appeal.
- Design-around may be too easy for competitors.
- Innovation may be too difficult for competitors to copy.
- Patent costs may outweigh the value of the patent.
- Patent-enforcement costs may outweigh costs of patent.
- Difficulty in identifying patent infringement.
- Profits from the invention may be short-lived and not patent dependent.
- The life of a product may be over before a patent is issued.
- Manufacturing efficiency and a dominant sales team may be more important.

PROMOTE

Get Your Invention Out There

Many of life's failures are people who did not realize how close they were to success when they gave up.

—Thomas Edison

Put the hours in. Doing anything worthwhile takes forever. Ninety percent of what separates successful people and failed people is time, effort, and stamina.

—Hugh MacLeod

ONCE YOU HAVE BEEN FORMALLY ADVISED BY THE PATENT office that you have a patent pending status (regular or provisional) you can approach licensees, joint venture companies, and manufacturers. Your goal is to get your innovation out there and profit. To attract either licensees, investors, or a partner, it helps to learn the art of promotion.

CARPET BOMBING AND STRATEGIC INSERTIONS—USE THE MEDIA TO PROMOTE AND MARKET YOUR INVENTION

The art of working the media is perhaps your most powerful tool. Tell people about your technology before it goes to market. There are thousands of editors sitting at their desks at this very moment wondering what to write about in their upcoming magazine or newspaper articles or what to cover on their radio or TV shows. Your challenge is to contact these editors and convince them to cover your invention in their publication or broadcast.

The Media Is Looking—Can They Find You?

Once you have practiced both the spoken and written word and bounced it off people you know and respect, you're ready to go for the big-time and approach the media. It's okay to be a media whore and a (polite) pain in the ass. Don't be afraid of getting hung up on.

In the media world, the big time is well represented by an excellent reference book, *Bacon's Publicity Checker,* that can be found in your local library reference section. It lists, by category, all the major magazines and newspapers in the country, along with the names and contact information for editors and writers. A useful online site is www.mondotimes.com. With a free basic membership you will have a personal media list to store ten of your favorite media outlets.

Before Sending a Press Release
- Study your targets and their areas of interest. Customize your press release.
- E-mail to advise that you are sending the press release and why they would be interested. Advise why their readers will care about your invention.
- Cut to the chase. Keep your press release short.
- Be honest when you hype your product.
- Send articles that others have written about you.
- Tell stories that add color to your main points.
- Send photos.
- If you can link your product to a well-known personality, be sure to mention it.

E-Mail
E-mail is ideal for sending flyers, newsletters, and promotional material. Although many people don't read their e-mail, they usually check to see where it came from and what it's about. Write short, to-the-point messages

in the subject lines. Mention the name of your product. Don't send spam or attachments without the recipient's permission.

In the old days, pre-Internet, I would generate a one-page elevator-pitch press release on RescueStreamer product technology, wait until after 11 PM, when telephone rates were cheapest, and then fax the one-page document to editors all over the world. One of my first international RescueStreamer distributors was in Japan and our entire (successful) relationship was conducted by fax. I never met anyone from this company in person or even talked to anyone on the phone for several years. This, of course, can be replaced by e-mail in the present day.

For every 100 faxes I sent, I might get two or three responses (some took follow-up telephone calls), but I finally got the *Miami Herald* to cover the RescueStreamer technology. Then, a few months later, I generated another elevator pitch and attached it to the *Miami Herald* article and said, "Look what the *Herald* is saying about the RescueStreamer product! Shouldn't you tell your readers about it?"

Within a few years, what started out as one newspaper article has now been published in every major magazine in the country—so much press that my scrapbook of press clippings is too big to carry to meetings. (It's about twelve inches thick!)

Send New Information and a New Hook

Note that each successive press release had a "hook," a new piece of information so compelling the media would have to write about it. (See the sample press release in the appendix.) Once I started getting military approvals, these became newsworthy items that leveraged more and more press coverage.

How to Make Contacts in Professional and Popular Journals, Radio, Television, and Print Media (or, Being a Polite Pain in the Ass)

Another strategy is to always go for national publications, or at least publications in states that you do not live in, first. It's easy to get local media coverage, but that is viewed as small-time (mom and pop). I prefer to go national first and then come back to local—it's bigger news! Note that in a local market you have to be careful about telling one magazine, TV station, or newspaper that their competitors have already covered the topic, as they commonly will not cover the same topic again.

People often don't read things you send them; therefore, a powerful picture, whether it is an actual picture in your print pitch or a picture you describe to them verbally, is worth its weight in gold. The text should be minimal, preferably in bullet-point format. Even if they only read a couple of bullet points, you may grab their attention.

The next important phase is the follow-up. Be a polite pain in the ass, but be persistent. It took me three years of calling back the editors at *Outside* magazine to finally get them to cover the technology. They kept blowing me off and saying, "call back in a few months." I took it literally, marked my calendar, and called them back in a few months. Each time, I would try to tell them what was new and compelling and which other magazines or TV shows had already covered the technology.

Use your credentials to get through to the right people. I use my "Dr." title (I have a PhD in science) to make me sound important. I also will leave a phone message simply stating my name, i.e., "please call Dr. Yonover at 808-xxxxxxx," with no other details. You'd be surprised how many people call me back, perhaps thinking they had some medical issue. Also, the fact that I have a PhD in science lends my pitch much more credibility as an inventor.

If you are an expert in a field related to your invention, make sure you mention that straight away, e.g., doctor, lawyer, military, teacher, work experience, or life experience related to the invention.

Getting Interviews: Their Loss Not Yours

Remember that any press is good press and that it's a numbers game. It took me hours and hundreds of faxes until I got a couple of nibbles that landed an article. Harden your skin and get used to rejection, and take the attitude that it is their loss not yours. I used to say to myself after getting shot down for the hundredth time, that they will regret not covering it when their competitor does—that is an attitude that's helpful in all aspects of life, from dating to inventing. Keep pitching and don't be afraid to fail!

This reminds me of high school days when girls wouldn't give me the time of day; however, now when I go back for reunions, they all want a piece of me. Payback is a bitch and you can get it with editors too! I worked on the new products editor for *Playboy* for years until finally he became a convert after seeing RescueStreamer technology in other high-profile magazines, including military magazines.

Unfortunately, the blurb in *Playboy* didn't include a photo, though I tried to get them to follow-up with the models draping the streamer around their bodies, but no dice. (Doesn't hurt to try.)

Using Blogs, Web Sites, Customer Forums, YouTube, Facebook, Twitter

Steven Yoder, author of the book *Get Slightly Famous*, thinks blogs are the best marketing invention ever created and having your own blog page and blogging others gives increasing competition in a big world. Short articles on your Web site and on other Web sites are a means of reaching your target market. Personal journal entries in your blog can double as marketing pieces. Blogs, Web sites, customer forums, and YouTube are all excellent marketing tools to help you and your product become "slightly famous."

Bill Clinton commented that when he first took office, only high-energy physicists had ever heard of the World Wide Web, but now even his cat has its own page.

Social multimedia sites like Flickr, YouTube, and Facebook have made publishing pictures and videos easy. YouTube is turning into an important marketing tool as it allows users to freely post entertaining videos that can include informative commercial pitches. Twitter supporters claim it operates faster than the blogosphere, enabling users to communicate with the masses and to send and receive a steady stream of virtual chatter and status updates.

Your Web site gives you access to Internet marketing, one of the most inexpensive and efficient modes of marketing available. Your site will attract customers from all over the world. Here, your target market can reach you. All you have to do is lure them to your door.

To Increase Traffic to Your Web Site

- Register the site with search engines (www.guerillapublicity.com/searchengines).
- Manually submit your Web site to search engines, following each search engine's submission guidelines. Make your Web site search-engine-optimization (SEO) friendly. (See below.)
- Create compelling keywords and add meta tags for keywords. For search relevance, content and title must contain some of the keywords. (See below.)
- Link to complementary sites. Contact webmasters of related sites to ask them to include links to your site. Add links to business partners, suppliers, customers, distributors, or related noncompetitive sites like industry guides. You can find which sites your competitors are linking to by using Netconcepts' free link-checker tool at www.netconcepts .com/linkcheck. Or type "Link:URL" in a Google search or a Yahoo! search.
- Make your Web site user friendly, easy to navigate, and effective. Test it on friends and ask them if they had problems with using the site.
- Make your site interesting and attractive and fast to download.
- Stay away from flash intros. Get to the message. You have only a few seconds to capture a user's attention.
- Don't enforce immediate registration. Wait until you've lured them in before you request (not enforce) registration. Put a good spam control on your system.
- Describe your product and its benefits. Give price, ordering, and shipping information. A contact phone number can bring good leads, but takes a lot of time.
- Update your Web site regularly. Have an "updated" line at the bottom and keep the date current. Ask people if they like your product and add their testimonials to the site. Ask if you can add their photos. Perhaps a photo of the customer trying and loving your product.

Mention upcoming trade shows and whether you have a booth. Offer free samples.

- Join discussion groups and forums interested in products similar to yours and mention your product and your invention and Web site. Ask for feedback. Many discussion groups allow FAQs (frequently asked questions) and information that's useful.
- You, the inventor, are an expert in your field. Add advice to the site. Write a list of frequently asked questions about an element of your product and the answers. Add a form so readers can e-mail questions to you.
- Add industry news to your Web site. Draw customers back with hot tips and information.
- Contact other inventors and manufacturers with related products and ask if they would like to be mentioned on, or have their Web site linked to, your Web site.
- When you receive an award or other favorable press, add the news to your site.
- Notify the press about your Web site on PR Newswire (www .prnewswire.com) and Business Wire (www.businesswire.com).

───── SUCCESS STORY ─────

Ron Bessems, Inventor of GIRDER

Software architect Ron Bessems, the inventor of GIRDER (a powerful and feature-rich computer- and home-automation application), sought a way to bring his product to the attention of potential customers.

Knowing that iPods were hot, he invented software to control GIRDER from the iPod and iPhone and created a short, four-minute video. In the first two minutes, he showed how cool it is to control your automated home from the iPhone, and in the second part of the video he detailed how easy it is to set up a system.

He then sent out press releases on the free Web site www.prlog.org that were picked up by the leading industry Web sites and used Digg (www.digg.com) a community-driven "linking" Web site to promote the video virally through user-generated buzz.

This is the best kind of advertisement, Bessems says, because it does not cost anything besides your time to create the content and is perceived quite positively by customers, as opposed to running expensive and usually not very effective banner ads. (See www.girder.nl.)

Keyword Research Tools

John Jantsch, veteran marketing coach and creator of the Duct Tape Marketing small-business marketing system, lists the following online keyword research tools that are very useful to help you create and refine your keywords for your Web site:

- Good Keywords: www.goodkeywords.com—free software.
- WordTracker: free light version: www.freekeywords.wordtracker .com. Pro version: www.wordtracker.com—a subscription-based and very powerful keyword tool.
- Google ADWords Keyword Tool: http://adwords.google.com/select/ Keyword/ToolExternal. Google's search suggestion tool based on Google's search date.
- Keyword Discovery: www.keyworddiscovery.com/search.html, an entire suite of search and keyword tools getting lots of praise.
- TopRank Blog by Lee Odden, www.toprankblog.com, has a nice list of search-engine-optimization and keyword-research tools.

By thoroughly researching the important keywords and phrases for your market before you even design your Web site, you may discover the most effective way to lay out your entire site to more effectively generate higher search-engine rankings.

Once you develop a list of keyword phrases to build your content around, you can create entire pages with the goal of ranking highly specific search terms. This type of high-quality, search-engine friendly content is one of the most effective search-engine-optimization practices available.

Google is currently offering Google Analytics, a very powerful tool that allows you to learn a great deal about the traffic coming to your site (www .google.com/analytics/sign_up.html).

TIP

Tip from Peter Kay, Inventor of Titan Key Software

How to Adjust Your Web Page to Improve Your Ranking with Google

Once you've assembled the keywords you want to "own" on Google and you've successfully begged other important sites to link to yours, all that's left is to make sure your Web pages have the right content.

Your page title, meta tags, and page copy must be consistent with your keywords for a high rating with Google. For our examples below, we'll use the keywords "extreme adventures in Hawaii," and we'll apply this to the Web site of a mythical company, Acme Tours.

Do your keywords appear in your page's title? In our example we might want a title such as "Extreme Adventure in Hawaii: Acme Tours Delivers!" Make sure your title is descriptive of what you do, includes your keywords, and uses fewer than fourteen words.

What about the fabled "meta tag," the hidden text within your Web page that search engines look at? If you want to find out your current meta tag usage, go to your home page, then, on your browser, click on view, then source. You'll get a page of some HTML code. Look for the words <META NAME=Keywords"> and you'll see your current list of key-words. In our example, we might want to include words such as extreme, adventure, Hawaii, tours, action, fun, excitement, surfing, paragliding, jet, ski, and jet ski. Most search engines will take only the first 1,000 words, so choose them carefully.

How often and how early do your keywords appear in your Web page's copy? Make use of your keywords in a natural language style throughout your page copy as often as possible, without being obviously repetitive. In our example, we might say, "If you're looking for the best extreme adventures in Hawaii, you've come to the right place."

Search Engine Optimization (SEO)

What if you go to Google and search using words that describe your company, and you're nowhere to be found? To get a higher listing on Google you need to perform what's known as "search engine optimization," or SEO. SEO is all about showing up at or near the top of the popular search engines. There are many companies (at least nationally) that can do this kind of work for you. Be prepared for potentially high costs and higher risk. The good news is that you can actually do much of the work yourself, and learn a lot about electronic marketing while you save money.

Your steps to SEO are:

1. Figure out what search phrase you want to optimize.
2. Find out what companies already have high rankings for those words.
3. Find out what those companies are doing.
4. Do it better.
5. Change your site and measure the results.

Figuring out what search phrases to optimize is all about thinking like a potential customer. The question you want to answer is "What would I search for if I knew what I wanted, but had no idea of the company that could provide it?" Come up with a list of various search phrases, review them with your trusted team, and pick the one that feels the most natural.

Now that you have your words, scope out the competition. Go to Google and type in your search phrase. The resulting list of Web sites that appears on the first page of Google is the one to beat.

SUCCESS STORY

Success from Nature—Bio Teams

Social insects such as ants, bees, and termites work together extremely efficiently using collaborative techniques that humans can imitate. In his book *Bio Teams: High Performance Teams Based on Nature's Most Successful Designs,* author Ken Thompson draws attention to the way nature's most successful teams work.

While traditional teams issue orders, bio teams provide "situational information" to team members who are trained to judge what they should do in the interests of the team: Ants use pheromones to transmit messages about predators and bees wiggle to tell other bees where the food is. Thompson says people can increase efficiency by broadcasting more whole-group communications and giving everyone access to information. So those who network on the Web could unknowingly be imitating nature's organizational bio teams.

SUMMARY

- Remember to keep your sense of humor, as writers and editors are people too. If you can get them to laugh, they will probably be more open to your suggestions. Ultimately, all of business (and life) comes down to relationships, so never burn a bridge unless you absolutely must.
- Send out press releases to the media.
- Practice pitching your invention, then do it.
- Google for conferences that offer a chance to speak to a potentially interested audience.
- Compile a mailing list of interested parties.
- Hold teleseminars with people on your mailing list.
- Use the Internet, have an interesting Web site, a MySpace page, and a personal blog, and use YouTube.

Find Investors: Targeted Strikes and Mobilizing the Troops

If your big plan depends on you suddenly being discovered by some big shot, your plan will probably fail. Nobody suddenly discovers anything. Things are made slowly and in pain.

—*Hugh MacLeod, author of* How To Be Creative

ONCE YOUR PITCH IS PERFECTED OVER THE PHONE AND through writing, you are ready to target people face-to-face in their comfort zones—in their clubs and, organizations, and other places where they gather.

IDENTIFY YOUR CUSTOMER GROUPS: CREATE INTEREST IN LOCAL ORGANIZATIONS AND CLUBS AND CONNECT WITH PROFESSIONAL SOCIETIES

Join clubs with interests that match your invention. For example, if you have a safety item for water sports, join a kayaking club. On the bigger scale, you can interact with general-interest social clubs, like Rotary, or go to local bars or hangouts with successful business-related people who could become investors or licensees. Medical doctors are famous for having a lot of money and looking for investments. Try to get a few minutes in front of them to pitch your product in a presentation or just in a one-on-one social setting.

THE ART OF THE PITCH, OR
HOW TO GIVE GOOD ELEVATOR PITCHES

The pitch is everything. You need to work at perfecting an excellent written and spoken pitch. In the venture capital world, professionals speak of your elevator pitch: the thirty seconds you may have when by chance you find yourself in an elevator with the president of a company that sells products similar to your invention or an editor who could write about your invention.

It takes a lot of practice. You have to grab someone's attention, get them interested, and get their contact information or get yours to them, all within a few seconds or minutes. Highlight the benefit of your technology and the benefit of covering it in their magazine or investing their dollars in it.

What's in It for Them?

This is the key question. It's one thing to say how great your invention is, but what do they care if it doesn't do anything for them?

I was once pitching a navy guy and making my key points on the RescueStreamer and how unlikely it is that you will be seen if the only target is your head or your white boat in a sea of whitecaps. He immediately saw the benefit of the streamer because he had been in that position—his sailboat had flipped over on a day cruise in the Hawaiian Islands. He immediately became a champion of the RescueStreamer and remains one to this day.

That is the ultimate, when a third party becomes a convert and strong supporter. That's when you sit back and let them as third parties preach the gospel—a similar effect to a magazine or newspaper story about your technology, since this is not a paid advertisement, but rather an impartial opinion of your technology!

Write an Outline of Critical Bullet Points about Your Invention

My personal style of pitching consists of minimal strategic statements or points that can be interjected into a conversation to convince the person to whom you're pitching (pitch target) that he should keep listening. I typically have a running list of bullet points written down that I continue to go over in my mind for either the planned or the chance pitch opportunity.

Remember, if you don't tell the world how great your product is, no one else will. Accentuate the positive. Eliminate the negative. I used the following bullet points to show the RescueStreamer technology:

ADVANTAGES:

- Continuous duration signal
- Attaches to personal vest or life raft—doesn't disperse
- No leakage
- Inert material—no chemicals, no battery
- Re-deploys
- Discretionary use
- Always works

Practice on your friends or family first; however, there is an element of familiarity with them you need to get rid of for a true elevator pitch. It's easier to talk to strangers at a gathering related to the technology, e.g., I pitch my RescueStreamer technology at Coast Guard or boating get-togethers.

The more strikingly visual your presentation is, the better. Try painting a picture with words, or use a PowerPoint presentation to show and tell.

With any pitch, it all comes down to connecting with the person or persons as quickly and efficiently as possible. This skill typically does not come prepackaged with your DNA. It has to be developed and fine-tuned.

The Fine Points of Pitching

The skill of pitching starts at the most basic level of being able to look someone in the eyes and listen as well as talk. I endeavor to look into someone's eyes or at a particular feature of their face if it is too hard to stay focused on the eyes. To be convincing, first and foremost you have to convince yourself that your invention is worth pursuing.

Size up your target before and during your pitch. As you present your pitch, watch their eyes and body language and be ready to adapt during the pitch. This acquired talent is priceless for all types of human interactions, and is the reason why person-to-person meetings (or telephone conversations) are the best if you can get them. The printed word and the spoken recording (voice mail) are one-sided and don't give you the opportunity to change on the fly, depending on how the target is reacting to your rap.

Practice Your Elevator Pitch

Once you have memorized your bullet points and have a general strategy, the best way to get better at pitching is to practice. Each time you pitch to a person (and it's best to try on family and friends first in case of a major blunder), try to critique your own performance and modify your next pitch accordingly. That way, by the time you have done it thousands of times, it becomes second nature. The pitch may only last thirty seconds. You pass a friend on the street and have thirty seconds before they have to go on their way. Your challenge is to translate your bullet-point message, modify, and ideally leave them wanting more—leading to an extended pitch or follow-up contact and subsequent more involved pitches.

How to Give Good Elevator Pitches

You are theoretically in an elevator and the person you have been waiting to meet gets in, e.g., CEO of a company that is an ideal candidate for licensing your invention. You have thirty seconds to make a pitch before the CEO gets out. If you are good, you can pique the person's interest enough to get invited to tell more or, send an e-mail, or to a follow-up (pitch) event.

The key to the pitch, especially the ones of short duration, is having powerful bullet points in your arsenal and being able to change bullets on the run. As you pitch the first bullet, read facial expressions and body language to gauge which bullet to break out next. Alternatively, you may detect that you have piqued the target's interest and he wants to speak or ask a question. The best sales technique is to let the customer/target speak—as long as you get your basic points across and it can lead to another meeting.

Sometimes you can get completely off track and talk about surf conditions; however, don't be discouraged, as making a human connection and not talking business may be better for building a relationship—and that is ultimately the name of the game. If you hit it off on a personal level, it will be much easier and even fun for the two of you to talk business at a later date. The key to a pitch like that is to make sure you get contact information: exchange business cards, get phone numbers, or ask if you can make a visit.

Sample Thirty-Second Elevator Pitch

Here is what I would say to an important person (e.g., a general, admiral, venture capitalist, or congressman) if I caught him for a thirty-second elevator ride:

Have you heard about my RescueStreamer technology?

The RescueStreamer is a patented orange streamer that gives you a forty-foot-long orange tail so people can see you if you are lost in the ocean, the desert, or snow—instead of your head being just a speck.

It's saved four lives to date—two in the military and two skin divers.

The newest version has automatic lights on it and works day and night, twenty-four seven!

Here is my business card. You can check it all out at www.RescueStreamer.com.

A Group Pitch

With a group pitch, I try to use the same principle: Make a connection with your audience. Combine eye contact with breaking out bullet points that connect with the audience. Never stop reading the eyes and body language of your pitch audience. If they appear to be getting bored by one portion of your pitch, go to another aspect that might be more interesting or intersperse a funny story. Once again—connect on a human level to build a relationship. Encourage questions and answers and try not to ever talk over people's heads. Try to translate complex issues into simple concepts so everyone in the pitch audience can be included at the same time.

Dress for the Job You Want

Unfortunately (and you can blame it on human nature) appearance is critical. You have likely heard the useful expressions "dress for success" and "don't dress for the job you have, but the job you want." I believe they both are true. Being a scientist, I always go one step further and analyze, predict, and carefully control my appearance relative to the specific target pitch person or audience.

If I'm going to a suit-and-tie international meeting, I go out of my way to first of all stay out of the sun for a couple of weeks before I go to the meeting (I don't want to rub it in the pitch audience's faces that I happen to reside on an island in the Pacific). I follow up the white pasty look with a fresh short haircut, exceptionally short if it's a military meeting I am going to attend. Finally, I break out a couple of suits, including one that is a bit more dressy than what the audience will be wearing. (Dress not for the job you want—no job!—but the role of a successful inventor whom people want to fund or buy technology from.)

If the meeting is more casual, I will try to match my attire with that of the pitch attendees. Even though I essentially try to match the audience's attire, I always try to dress a notch more formal. I want to show the pitch targets that I am serious about the pitch and possibility of working together in some capacity.

The same goes for speech and mannerisms. If I meet a pitch person at the beach, I talk in a more subdued manner (at least I try to contain myself), peppering in some more impressive bullet points with the more folksy conversation points. I am careful in any situation, from casual to formal, not to make any obvious verbal mistakes or misuse words. It is an obvious flag to me when someone uses a word in the wrong context or makes up a word. Even a minor mistake sends a negative message. Whether conscious or unconscious, it affects the relationship you are trying to build, so choose your words and actions carefully.

This has proven to be a major challenge for me since sometimes my mouth starts speaking faster than my mind can filter and check the output. One thing is for sure: If the pitch target is a major one, I am extremely careful with the words that come out of my mouth.

Also a major advantage of a person-to-person pitch is you can perform damage control if you make a verbal or social error (like disrespecting an aspect of the pitch target's life, business, or friends without meaning to). A quick read of the pitch target's face and mannerisms can tip off that they are pissed or uncomfortable with something you just said. When this happens, you have a split second to perform some damage control on the inappropriate statement or you can change the subject to go to a safer subject. Of course, avoid discussion about politics, religion, and other sensitive subjects.

National Trade Shows

The granddaddy of the "give good elevator" pitch event is the trade show. There is no better place to connect with the power players in an industry sector. All the players are there and it's essentially an even playing field. Everyone is dressed the same, out of their offices, and open to new concepts and meeting new people.

Look for national and international trade shows that encompass the field in which your invention fits. You can find out about trade shows in any industry on www.tsnn.com.

A first step is to pick a trade show and attend as a simple "attendee." Get the program and plot the booths and people you want to visit and pitch. Get familiar with the names of people you want to pitch and get ready to do some heavy name-tag scanning. Ideally (if you are protected by a patent pending) have your prototype or a detailed sketch and one-page description. And bring business cards to leave behind. Have your business card describe you the way you want others to perceive you.

Walking the Floor

The low-budget "bottom-feeder" method to maximize your time at a trade show is to simply register and "walk the floor," avoiding the high cost of your own booth. Unfortunately, the challenge in this method is that the pitch targets you need to talk to are also walking the floor and stopping at people's booths. You have to be more aggressive and intrusive and loiter around other people's booths and essentially hijack their pitch targets.

Loiter around a booth that has a lot of traffic that could be related to your technology and, once again, be sociable and persistent. Although you may get a sore neck reading name tags, I found it to be the most efficient way to identify and engage with people you have targeted. Wait your turn and then pounce (in a nice way) with your elevator pitch. Don't be afraid to be rejected—it only takes one good connection to make the whole effort worthwhile. You never know where you will make the connection; it could be in the bathroom or the lunchroom. Keep pitching, and if people shut you down, try to get some useful information from them as they shut you down—ask them their opinion on where to go next or if they know any people or companies that may be interested.

You can also meet your target pitches at social events, meals, and other locations at the venue. Some of my best work over the years was performed at the social events. Your target pitches usually have their guard lowered when they are at a party and are slightly or considerably liquored up. Once again, if you just meet them and have a primarily social conversation, you can leverage that the next day at the "work" part of the trade show.

I am sensitive to talking shop at a party, so I usually talk about non-business topics, but I always remember to interject a few bullet points they will remember when I encounter them again at the formal part of the trade show.

Another useful strategy for walking the floor at a trade show is to essentially have your "mini-booth" show in your pocket or briefcase. If your invention is small enough, carry it on your person. If not, carry photographs of your creation, especially if you have shots of the invention doing its thing.

When meeting a pitch target on the floor or in a booth, I've come up with a phrase I would patent if I could. It works best when you are standing at your own booth with pitch targets strolling by (or any retail sales situation). Over the years, I have observed people at booths (or stores) say the two kiss-of-death lines: "Can I help you?" or "Hello!" Both lines appear to be useful or fairly innocuous, but they usually result in a one-or two-word response, with the pitch target moving on to the next booth. They are not engaging questions, and no one likes to admit that they need help.

The line that works every time is "Have you seen this?" If the target says "yes," you proceed to tell them about the new version or update them. If the target says "no," then you tell them about the invention or technology. Either way, the target is looking at what you are showing them and thinking about whether or not they have ever seen it. And then the stage is set for your pitch to be delivered.

Hosting Your Trade Show Booth

If you're fortunate enough to make it to the hosting-your-own-booth stage, consider the booth as an extension of your self. You need to dress up the booth the way you would dress yourself to match your targets—not too casual or too formal. I try to make my booth inviting to the targets as they window-shop the multitude of booths. An open, uncluttered booth delivering a simple visual message is best. I like to have a minimum of furniture in the booth and an inviting photo display on the back wall that lures the targets into the booth so I can spend some quality elevator time with them.

I dislike standard-sized chairs or tables because they put you at a lower level than the targets walking the floor. Also, a target may feel bad if they make you get up from your seat to talk to them. To avoid this, I prefer tall stools to the side of the booth so even if you are seated and resting your feet, you are eye-to-eye with the targets and can easily make the transition from seated to standing. You want to treat your booth like your living room, a place where targets can feel comfortable, peruse your visual displays, and engage in discussion.

Working a booth is a physical and mental challenge. I actually try to train before I go to a trade show. I practice by standing on my feet for extended periods of time and, depending on the meeting makeup, I may even practice drinking beers and partying, since that is a skill you will need to work the socials at night! Remember, when people are having fun, they are most open to thinking outside the box and to new ideas.

If you can afford it and have your product in a mature state (patent pending with prototype, artwork, etc.), having your own booth is the best way to get people to check your product out. However, if you are not ready, it could backfire by looking amateurish. In that case, it's better to have your (patent pending) prototype (or a drawing or picture of it) in your briefcase and simply walk the floor and look for targets.

One further note about trade shows: Be prepared to transition from your phone persona to your "in-person" persona. I once had a few military guys flip out after they met me—they completely let their formal guard down. They said because of my "Dr." title and my deep phone voice, they thought I would look like Colonel Sanders—old and white-haired with glasses. They were surprised at how I looked and came across. I put them at ease and told them I only use that title to break the doors down and penetrate security.

Since relationships are so critical in business, at trade shows, and at other professional meetings, make sure you don't miss the after-hours social functions—this is where the real business and relationships get cast in stone. Once at a formal social at a military meeting, I was standing next to some elderly women. We struck up a lively conversation. It turned out their husbands were major power brokers and they ended up introducing me to them. Although I didn't intend it that way, it pays to build relationships with everyone in your adopted community at your specific professional trade shows.

TIP

From Barbara Pitts and Mary Sarao
Inventors of Ghostline

The inventors of Ghostline, on their very helpful Web site www .asktheinventors.com, recommend joining one of the associations that sponsor the trade shows. You can probably join as an associate member for a reasonable fee.

Attend the show and walk around making notes on the companies that look like potential licensees of your product. On the last day of the show, go to the booths you identified as potential licensees and give them your business card. Say you have a product that fits right in with their line of products and you would like to call them to set up an appointment to show it. Do not tell them what the product is; simply say that it solves such-and-such a problem.

Get a business card from any helpful person you talk to and in a week call and remind them that you met them at the trade show. When you say you are a member of the association, you will be treated as a professional familiar with their industry. Make an appointment and go and pitch your product.

TIP

From Susan Wyshynski, Coinventor of the Mandala System and VTV

Wyshynski says if your idea is exciting or a "draw," some trade shows might even help with your costs, such as giving you a free booth—it doesn't hurt to ask. Call the trade show event organizer and let them know how good your invention is. Ask if there is any way they could help you out with any of the expenses or to reduce costs. Call one of the big guys with big booths at the show and let them know that you have a really good thing that would be a draw in their booth. In this win-win situation, Wyshynski and her co-inventors were paid $5,000 to set up their invention in the booth. The chance of playing in a virtual "Hockey Hall of Fame" brought customers into the booth, attracting buyers for the booth holder and for the inventors.

From David Parrish, Engineer and Innovator, on the Marketing of the Fish 'n Flush Aquarium Toilet Tank

The Fish 'n Flush is a toilet tank that doubles as an aquarium with a unique twist—when you flush, it looks like all the water is draining out of the tank and the fish are going down. But they're not—it's an optical illusion. The fish are actually safe in an outer aquarium made of highly polished plastic. What you see draining is the toilet water in a separate tank behind the aquarium. The tank refills and the fish continue happily swimming. Great for potty training!

When the Fish 'n Flush hit the market, the creators had trouble figuring out what market to concentrate on—what buyer, when they see a toilet tank full of fish, just has to have it? Despite some initial hiccups (not from the fish), the product—first created as a gimmick—has been very successful.

Buyers have included a restaurateur in Paris and a fisherman from Georgia who populates his tank with baby bass. Creator Richard Quintana has also shipped orders to Korea, Kuwait, Hungary, and the Netherlands. "And it's big in Canada," says David Parrish, Quintana's colleague.

David Parrish gives his advice on marketing and trade shows:

"To better understand the success of the Fish 'n Flush, a brief history of its evolution from test-lab fun to a viable product is necessary. The Fish 'n Flush was a simple idea imagined one afternoon in 2005 in the AquaOne testing lab during a design review for our core product, the H2Orb, a device that stops toilet overflows. The AquaOne team (Richard Quintana, CEO; David Parrish, COO; Brian Reel, director of sales and marketing; David Millar, mechanical designer; and Alfonso Cano, production supervisor) was thinking of ways to attract attention at a trade show when the idea of putting goldfish in the clear acrylic H2Orb demo tank was laughingly proposed. The next day, our mechanical designer had a design for a two-piece tank and less than ten days later a working prototype.

"The tank was left in the shipping box for several months at one of our satellite offices. As a small joke, we 'unveiled' the unit at a show-and-tell meeting with the president, vice president of engineering, and vice president of marketing of a $13.7 billion-dollar building-supply company. The Fish 'n Flush idea was a great icebreaker and after the unveiling we left the fully operational tank in our showroom. Several weeks later we received a phone call from a representative of one of the top sensor companies in the world, one of our vendors, who had been by the shop for a sales visit. They wanted to offer us a free booth space at a security trade show in Las Vegas; 'Bring the Fish 'n Flush,' they said. April 2006 was the first public launch of the Fish 'n Flush Toilet Tank Aquarium. A Fish 'n Flush prototype brought a huge crowd to the booth along with unexpected media attention. We reintroduced the prototype at the Kitchen and Bath International Show (KBIS). While the Fish 'n Flush did extremely well at KBIS as far as media attention, it did not bring tremendous sales. Based on the strong interest we had received at several trade shows where the Fish 'n Flush was an 'attention getter' gimmick and not a serious product, we jumped into several product shows with both feet.

"For example, the Fish 'n Flush, being a pet-based product, or so we thought, led us to Super Zoo, a pet-based trade show in San Diego. Serious aquarium aficionados do not do small fish tanks. We had lots of interest and not one sale. The product placement of the Fish 'n Flush was more decorator based than pet industry based. If we had worked out consignment deals with several local pet shops, home stores, interior design stores, etc., we could have targeted our marketing and advertising dollars and our time more effectively. We could also have attended local and regional shows and found our niche market.

"Bottom line is that I believe that trade shows can be effective, but pick and choose wisely. Attending a large trade show like KBIS can work several ways; you can get lost in the crowd, you can get discovered, or you can go home discouraged and beat when in reality you were just at the wrong place. Our team all agreed later that KBIS was a good move, but we should have done more homework beforehand to maximize the exposure." (See www.fishnflush.com.)

Invention Competitions and Science Fairs

Colleges, schools, organizations, and companies all encourage invention and young inventors. Some ask for specific types of inventions such as ones that will benefit the environment, increase robotic expertise, produce new toys or games, or even increase the use of Bubble-Wrap. Science fairs encourage students to conduct experiments and research topics. Invention competitions ask for new inventions. Some competitions are open to individuals and some to teams. All these competitions offer publicity and the opportunity to network with people who could help fund, manufacture, license, or promote your invention. Many offer sizable cash prizes and college scholarships.

NASA encourages robot inventors with NASA's Robotics Alliance Project, and robot and botball competitions that bring students, engineers, private organizations and government resources across the country together with the goal of increasing the nation's knowledge of robotics (www.botball.org).

The FIRST (For Inspiration and Recognition of Science and Technology) program and FIRST Robotics competitions inspire young people to be science and technology leaders. Participants engage in exciting mentor-based programs that build science, engineering, and technological skills, inspire innovation, and increase knowledge and life skills. The founder and driving force behind the program is Dean Kamen, the incredibly successful inventor of the Segway PT. See www.usfirst.org for details of FIRST Robotics competitions, FIRST Tech Challenge, and FIRST Lego League.

The Defense Advanced Research Projects Agency (DARPA) holds a robotic car race with a $2 million grand prize. The driverless cars are not remotely controlled. Instead they use sensors and software to "figure out" the mock urban military supply missions course for themselves. The top two teams in the Pentagon-sponsored race both used Applamix's Position and Orientation Systems for Land Vehicles (POSLV), an inertial/GPS mobile mapping technology.

In 2007, a robotic vehicle developed by Carnegie Mellon University professors and students won the $2 million grand prize. Stanford University took the $1 million second prize. The competitions are designed to spur research into technologies that eliminate the need to send military troops into dangerous battlefields.

The importance of increasing the nation's robotics expertise is enormous and offers many opportunities for inventors and scientists. To design future generations of robots "able to think on their feet" and even scarily more powerful than humans, areas being researched include machine vision, three-dimensional mapping, and automated systems for mobility.

How to Succeed in Science Fair/Invention Competitions

- Decide which contest and category is best for your invention.
- Each competition is different, so make sure you read the guidelines carefully.

- Make a list of submission requirements.
- Know the deadline for the entry.
- Work on a good display board or poster with clear illustrations and clear descriptions of your invention. Give a quick overview of the question you asked, the method you used, the results you got, and the conclusion you came to.
- Mount an eye-catching title in large letters on the display board.
- Draw charts, diagrams, or illustrations to explain your question, methods, and results. Mount items on colored background card-board.
- Your notebook with details of your inventing process is an important part of your presentation. Display it with your poster for those who want to know more about your project.
- Demonstration materials illustrating a scientific principle, equip-ment, or materials used will make an exhibit more interesting. Such materials should be placed in front of your backdrop display on a display table.
- If appropriate and feasible, display photographs of people using your invention.
- Keep an emergency kit with scissors, glue, staples, markers, etc., handy.
- Practice your oral presentation. Prepare a bullet-point summary of your invention and the reasons why you invented it and why it is useful and new.
- If your experiment involves animals, dangerous chemicals, or valu-able equipment, take photographs to illustrate your work instead. Exhibits will be left in the hall overnight and examined by many other students and their families. Don't risk damage or harm to your-self or others.

The following success stories illustrate how invention competitions and science fairs help inventors achieve recognition and assistance in developing their products.

Atlas Powered Rope Ascender

The Atlas Powered Rope Ascender was designed to address a challenge presented by the U.S. Army in the Soldier Design Competition, hosted by the Institute of Soldier Nanotechnologies (ISN) at MIT. Challenged to create a powered rope ascender suitable for use by the army, Team Atlas built the first working prototype of the Atlas Ascender, and won the competition's SAIC award. The impressive device is a machine powerful enough to reverse-rappel a man up a wall at ten feet per second and small enough to fit in a backpack. The Atlas Powered Rope Ascender was also winner of the PopSci Invention Award 2007.

Inventors Nate Ball, Tim Fofonoff, Bryan Schmid, and Dan Walker designed the machine in just three months for $700. The rider wraps the rope around a motor-driven grooved spindle. As the spindle turns, the rope pulls tight, preventing it from slipping while small wheels guide it through. The group was helped by an MIT professor who founded the company that produced the lithium-ion batteries used in the device. The team has formed Atlas Devices LLC to focus on the advanced development and commercialization of the Atlas Ascender and to provide its benefit to the military, law enforcement, and rescue workers as quickly and effectively as possible. Starting with the Atlas Ascender and going far beyond, they are dedicated to developing creative technologies that benefit and enhance human capabilities in extreme situations. Already the army has placed several orders for the Atlas Ascender. It could be used in urban warfare, to evacuate casualties from city streets to rooftops, in military rescue operations, or to zip in and out of terrorist-harboring caves. (See www.atlasdevices.com.)

Illuminated Nut Driver, Kristin A. Hrabar

When Kristin was in third grade, her father, Frank, asked her to hold a flashlight for him while he worked on the family's clothes dryer. Kristin had a short attention span—she'd been classified with a learning disability and attention deficit disorder—and she quickly grew tired of holding the flashlight and started flashing it around. Her father couldn't see what he was doing. He shouted at Kristin to keep the flashlight still. Then nine-year-old Kristen told her father he wouldn't need her to hold a flashlight if the tool had a built-in flashlight.

"Great idea," her father said. "You can make it your science fair project." A few days later, Kristin created a prototype of the Illuminated Nut Driver with a battery-powered pen light, a straw, and a nut socket.

Kristin's final product has two LCD lamps surrounding the outside of the tool while a laser lamp sends light through the center of the fully hollow nut driver shaft. A composite material for the tool shaft makes it safe for use around electricity.

Kristin received a utility patent in 1998. The following year she won the 25th Annual U.S. Patent and Trademark Office Expo. She was named a Walt Disney and McDonald's Millennium Dreamer Ambassador to the World in 2000. Kristin received a second patent in 2001, appeared on *Inventor's Showdown* on the Discovery Channel, and was named "Young Entrepreneur of the Year" by Partnership for America's Future, among many other honors and awards. See Kristin's very successful Illuminated Nut Driver and other products on her Web site at www.lightsabertools.com.

GPS Robot Warren Jackson

Better data is on its way for climate-change researchers thanks to an invention by university engineering student Warren Jackson and a team of students from the Weiss Tech House at the University of Pennsylvania.

Every year the National Weather Service launches about 80,000 weather balloons containing expensive devices called radiosondes, and they lose most of the radiosondes, a loss that costs about $10 million a year. Warren Jackson developed the idea for an autonomous global positioning system–based robot in 2006 and received a summer internship to work on it. He recruited a university team experienced in robotic aerial vehicles and they formed Radiosonde Recovery, the first-place winner at the 2007 PennVention competition.

The PennVention competition begins each year in early November and culminates at the Final Invention Fair in April. Participants receive business mentoring services and a chance to win a portion of $60,000 in cash and in-kind services. Past PennVention winners have gone on to secure patents, incorporate businesses, sell their products on QVC, solicit celebrity endorsements, and contract with the U.S. government. (See more at www.tech-house.upenn.edu.)

Sign Language Translator Glove

Ryan Patterson, the inventor of the American Sign Language Translator, displayed an early interest in electronics and engineering, inventing robots while still in school.

Watching a group of deaf people in a restaurant struggling to tell the cashier what they wanted to eat, he decided there had to be some simple way to make everyday communications easier for deaf or nonspeaking individuals. Working with a mentor, John McConnell, a retired physicist, the seventeen-year-old Patterson developed a prototype for a sign language translator, a glove that works by sensing the hand movements of the American Sign Language and sending the data to a device that displays the words on a small screen.

Patterson obtained a provisional patent on the system and has since made improvements such as incorporating speech dictation software. His remarkable invention earned him the Grand Award at the 2001 Intel International Science and Engineering Fair, and top honors in Intel's 2002 Science Talent Search, where he was awarded a $100,000 scholarship. He was also first-place winner in the individual category at the 2001 Siemens Westinghouse Science and Technology Competition, which earned him another $100,000. He received a full-ride Boettcher Foundation scholarship. This young inventor's invention and other academic achievements helped him earn more than $400,000 in scholarship money. After earning a BS in electrical engineering and computer engineering in 2006, Patterson continues to work on projects such as an indoor global positioning system that might aid people with cognitive disabilities by allowing their loved ones to monitor their movements from afar.

SUMMARY

- Find investors. Target potentially interested people in social clubs, local organizations, professional societies, and sports clubs.
- Attend trade shows.
- Work on perfecting your pitch, on the road and at trade shows.
- Enter invention competitions.
- Enter science fairs.

Protect Yourself from Fraudulent Invention-Promotion Companies

Take the long road 'round the forest while the wolf is in the wood.
    —*Practical Pig,* The Big Bad Wolf *(1934)*

MANY INVENTION COMPANIES OFFER TO HELP INVENTORS get their products in front of potential buyers for a fee. Some companies are trustworthy, but many are not, and there are myriad horror stories of inventors being scammed. How do you protect yourself from scammers?

Jim Lowrance, a successful inventor who was scammed early on in his inventing career, has the following advice:

- Check the credentials of any invention company as thoroughly as you can.
- Ask for references from other inventors the company has successfully acted for.
- Get a statement of achievements and check out success stories.
- Look for an invention firm that charges only an expense fee, their main profits being made from the percentages of royalties they have secured for inventors.

- Take a lot of time before making your decision.
- Check out the U.S. Patent and Trademark Office's public forum for invention promoters and promotion firm complaints and responses, as well as disclosure rules. And check out the USPTO's FAQs for inventors.

AMERICAN INVENTORS PROTECTION ACT

The large number of fraudulent invention-related companies and resulting legal activity caught the attention of Congress in the late 1990s, and lawmakers approved the American Inventors Protection Act of 1999. The act provides that the U.S. Patent and Trademark Office keep a complaint database for consumer reference. The American Inventors Protection Act gives you certain rights when dealing with invention promoters. Before an invention promoter can enter into a contract with you, it must disclose the following information about its business practices during the past five years:

- How many inventions it has evaluated
- How many of those inventions got positive or negative evaluations
- Its total number of customers
- How many of those customers received a net profit from the promoter's services
- How many of those customers have licensed their inventions due to the promoter's services

This information can help you determine how selective the promoter has been in deciding which inventions it promotes and how successful the promoter has been. Invention promoters also must give you the names and addresses of all invention-promotion companies they have been affiliated with over the past ten years. Of course, the fraudulent companies reveal this data at the last possible moment.

Use this information to determine whether the company you are considering doing business with has been subject to complaints or legal action. You can also check the company by calling the U.S. Patent and

Trademark Office, the Better Business Bureau, the Consumer Protection Agency, and the attorney general in your state or city, and in the state or city where the company is headquartered.

The Rod Floater

In 1984, Jim Lowrance and his brother co-invented a fishing tackle accessory called the Rod Floater, a flotation device to keep rod-and-reel combos from sinking. They found ads promising inventor assistance in magazines and chose the best-looking one, ordering an information package. The first package promised success. The next basic information package cost just under $400.

The invention company praised the product and advised that it would negotiate a contract with a manufacturer that could secure the inventors a 5 percent royalty.

This next step cost the inventors a further $4,500. Two years later the invention company was still making promises and referring to deals with companies with global-sounding names. Later, checking the companies, Lowrance found that the main supposedly interested company was nonexistent. None of the other companies supposedly contacted by the invention company had ever been called, and none had knowledge of the invention.

After a further two years of promises and waiting, Lowrance and his brother decided to market the invention on their own. They approached local businessmen in their hometown who were honest and reputable. Lowrance made a basic pitch about having a patentable invention and two of the businessmen came to a meeting at his home. He got one major investor who agreed to finance the project, up to the point of having a packaged, patent pending product set up for production, for 40 percent of the profit made. Afterward, another major investor came aboard and he invested a fairly large sum, in exchange for 20 percent ownership, and yet a third investor secured 5 percent ownership for his investment. Lowrance retained 35 percent ownership.

They immediately experienced success getting the Rod Floater into regional Wal-Mart stores, then Bass Pro Shops and other reputable retail outlets. They also secured distributors, sales reps, and TV marketing and landed a promotional deal with a large oil company who used the Rod Floater as a promo giveaway by placing one in their cases of outboard motor oil. Now that the product is under a license agreement, payments remain steady.

Lowrance says if he'd known the product was going to be licensed a few years down the road, he likely would have kept more of the ownership, but that's one of the risks of getting investors involved. Part of the satisfaction of success with an invention is seeing how well it is doing and its longevity on the market.

Lowrance also invented outdoor pouch products that he sold outright to a company in Illinois. He has written a number of articles to help other inventors which can be accessed on the Web at www.inventorspot.com. He advises, "It's an incredibly difficult thing to go through, but if you're ambitious enough it'll work for you."

PROFIT

Licensing

You miss 100 percent of the shots you never take.

—Wayne Gretzky

Never let an inventor run a company . . . you can't get him to stop tinkering and bring something to market.

—E. F. Schumacher, economist

THERE ARE TWO MAIN OPTIONS WHEN TRYING TO DECIDE how to make your invention succeed:

1. Licensing to a manufacturer
2. Manufacturing your product

LICENSING

Licensing your product to a manufacturer is usually the most feasible path. Licensing is an exciting alternative to going into business as it helps defray the high costs of development, manufacturing, and marketing. It is frequently the fastest way to launch an innovation and earn profits.

What is Licensing?

The inventor spends the time and money to hire an experienced patent attorney to obtain a patent or provisional patent. This gives the inventor valuable intellectual property and the exclusive right to manufacture, use, or sell products falling under the scope of the patent. The license agreement transfers that right to the licensee.

Usually, the inventor sells the intellectual property (IP) to a manufacturer for either a onetime fee or royalty, or a combination of both. As an inventor, you probably want to keep inventing more stuff, not run a business. (See *The E-Myth Revisited*, a great book dispelling the false belief that people want to run their own businesses. The author, Michael E. Gerber, states that you should build a business to sell it, not to run it.)

As the inventor who licenses an invention, you may need to locate and work with a manufacturer to build more advanced prototypes or even products for the market; however, the end result should still be selling the IP to a group that can run and grow the business and send you the coveted royalty check.

Licensing should be seen as gaining the necessary funding to launch your invention without personally being at risk. As the manufacturer has assumed the risk, when licensing a technology or product that is unproven in the field they will probably pay a conservative price, possibly a lot lower than the inventor thinks the product is worth. Still, licensing is often, in my opinion, the best way to go. Sell at the right price so it goes quickly; nobody likes to buy an old invention.

To license your invention you need to:

- Locate manufacturers who could be interested in your product.
- Prepare marketing material.
- Check with a patent attorney regarding confidentiality issues.
- Submit marketing material to manufacturers.
- Find an experienced attorney to negotiate a licensing agreement.

Locating Licensees and Manufacturers

Search local stores like Wal-Mart for the names of manufacturers with products similar to yours. Check the Yellow Pages. Trade shows. Check government agencies and magazine advertisements. Local libraries have resource books like the *Thomas Register* and the *American Business Directory*. Internet databases are useful too. Try www.kellysearch.com, www.housewarebuyers .com, www.industrynet.com, and www.kompass.com.

Trade shows are good places to locate a potential licensee. Google has links to lots of trade shows (type in your category of invention and "trade shows"). At these shows the manufacturers of your type of product will have booths and will be selling to the retailers who attend the show to order products for their stores.

Finding small manufacturing companies and marketing firms wishing to expand may work better than trying to negotiate with large companies. When contacting a company, it's often better to pitch your ideas to someone in sales or managers in marketing and not top-level executives who are too busy.

Network

Meet experts at the Licensing Executive Society, the Product Development and Management Association, and the American Intellectual Lawyers Association. Talk to other inventors at invention shows. Find a product that is a good companion product to yours and locate the inventor by finding the patent number and pulling the patent at the Patent Depository Library. Talk to the successful inventor. How did he find a licensee? He may be willing to refer you to his licensee. Gather information at conventions. Small-business associations may have surveys on manufacturing and marketing companies in your area who are interested in new products. See also Weddle's Association Directory, which has forty-five or so manufacturing associations listed under the category Industrial/Manufacturing.

Once you have compiled a list of possible manufacturers and licensees, do your homework. Research the company well so you understand its business, products, and customers. Make sure to check the marketing and sales expertise of the potential licensee. Does the licensee have distribution channels already set up?

Licensing—A Strategy for Profits, a book commissioned by the Licensing Executive Society, has case studies and an overview of the process of licensing. Very few innovations are overnight successes. Sometimes the difference between the ones that make real money and those that don't is how seriously the inventor and the licensee chase that pot of gold.

The National Inventor Fraud Center Web site, at www.inventorfraud .com, has advice and lists of companies seeking new products.

Presenting Your Product to Potential Licensees

The licensee, usually a manufacturer, is your customer and not the final purchaser of the product. Keep in mind that they will be selling the product to their customers. Your job is to sell them on why your product will be beneficial to those customers. Companies want a proven product that will sell.

Written Presentation

Before you meet with the potential licensee, prepare a written presentation that explains what your invention is and why it will be beneficial to the potential licensee's customers. Prepare two copies for each company you approach.

The written presentation should anticipate all their questions. (See bulleted points for presentation below.) Explain what your product is, how it works, why it is needed, what problem it solves, how much it will cost to manufacture, and why it would be a profitable item. If you are able to meet the potential licensee in person, give them the written presentation after you have completed the oral presentation. An excellent written presentation has the potential of selling your product to those who do not attend the meeting rather than relying on your contact to sell your product for you. Get the presentation bound in a professional-looking binder.

Your next step is to contact the new-products divisions of each of the best twenty companies by phone. Call yourself a product developer and not an inventor, and use a company name if you have one. Ask for a specific person to mail the marketing letter to.

Send brief, professional letters tailored to the specific companies asking if they are interested in purchasing or licensing the patent rights to your invention. Do not give a detailed description of your product. Just describe how it solves a common problem in a simple and inexpensive way. Request a face-to-face meeting and schedule an appointment. (See sample marketing letter in the Appendix.)

If your invention has a provisional patent or is patent pending, you may want to include some of your professional patent drawings with the marketing letter. Before sending the letters or patent drawings and before disclosing any detailed invention information, seek the advice of an experienced patent attorney regarding confidentiality issues.

You could also attempt to have the manufacturer sign a confidentiality agreement prior to disclosing your full invention. Unfortunately, many will not agree and some will want you to sign their disclosure form that states you cannot make claims against them. You may have to do it their way if you still want to show it. (See sample nondisclosure agreements in the Appendix.)

In your written presentation, and when describing your invention or new product to the potential licensee, consider the following guidelines:

- How does the product fit in with their product line? Does it allow them to offer their customers something new and different? Does it enable the licensee company to enter a new market? How does it solve a market need? Find out all you can about the company's goals and tailor your pitch to fit in with what they are already doing.
- What is the manufacturing cost-to-retail ratio? Preferably it should be no more than 1 to 4. This means if the product costs $1 to manufacture, the licensee has to be able to sell it for no less than $4.
- Are you already manufacturing and selling the product? A proven sales history reduces risk.
- Do you have a working prototype showing that your product works?
- Is your product patented or patent pending?

If you are not meeting with the company in person, send a personal letter with the presentation. In the letter, explain who you are, and why and how you invented your great product. Send the presentation in FedEx. If you are dealing with a large manufacturer with a research and development department, submit your product through them; otherwise, submit it to the president or vice president of the company.

If you have not heard anything from a company after approximately four weeks, you should contact the company by telephone to see if they are interested in purchasing or licensing the patent rights to your invention.

Science or Technology

Using an intermediary is an option when trying to get your science or technology idea in front of corporate buyers. Corporations hire firms like Yet2.com to license products for their science and technology needs, outsourcing part of their research and development. Inventions must have proven value before they will be considered; Yet2.com will screen and make sure the idea is a good fit before making a direct connection, which usually results in a confidentiality agreement. All fees are paid by the buyers, including companies like Honeywell, Microsoft, and Siemens.

How to Sell Your Invention—Let Them Hold It, Then Take It Away

When trying to close a deal with investors or licensees, you have to sell them the dream—your vision of where this technology is going. My vision and the vision I licensed for the RescueStreamer technology have come true: RescueStreamer is today a standard piece of lifesaving equipment in all military, commercial, and civilian sectors—making money and saving lives at the same time!

To sell this vision, I used a successful technique I learned back in preschool—if you want to play with someone, you share your toys. You let them touch and feel the toy; however, you take it away from them if they are not committed to playing fair or sharing in the fun. Same thing in business—you let a potential investor or licensee hold your product, physically and emotionally, and then you take it away from them if they are not going to follow through with a license or investment.

When they are holding it in their hands, sell them on the dream and the vision and how they will benefit from it. Get their eyes to glaze over as they share your vision. Then, take it away from them and begin discussions of a legal agreement. They will feel temporary ownership and ideally will miss it when it is taken away from them and they know they could have shared in the vision.

Negotiate an Agreement

Once you have found a manufacturer or licensee interested in your product, you need an attorney to help negotiate an agreement. Pay attention to how the royalties will be paid. There are two ways you can profit: flat fee or royalty. Flat fees will usually be low as they will be based on pessimistic sales projections. Many inventors who sold outright later wished they had chosen royalties. This way you and the manufacturer or licensee win or lose together. Points to consider and discuss are:

- What will my royalty be?
- What will be my up-front money?
- How can I keep a company from sitting on my idea and not commercializing it?

Royalties are governed by expected profits: An average royalty is 6 percent; higher than 7 percent can only be expected if gross profit margins are high. Royalties will be closer to 3 percent if lower profit margins are expected.

Most up-front money ends up lowering the royalty rate. Many licensees are not eager to pay an up-front fee. So try for up-front money, but don't let it kill the deal.

There should be a minimum royalty clause in the licensing agreement. If these royalties are not met, the inventor should have the right to terminate the agreement and find another licensee or renegotiate the contract. (See sample exclusive license agreement in the Appendix.)

The licensing agreement should contain a clause to allow the inventor to audit the licensee's records once a year. Further clauses should grant indemnification to the inventor, who should not be responsible in the event of an infringement suit. A clause should include the need for a product liability insurance policy.

The agreement should contain a clause that the licensee must actively promote your invention or else its rights are terminated. (See sample license agreement in the Appendix.)

Summary of a License Agreement

- Inventor grants the licensee the rights to manufacture, use, and sell products falling under the scope of the patent and intellectual property.
- Establishes geographic or market limitations.
- The licensee pays royalties: a minimum royalty clause, payment terms, and up-front royalty payments.
- Licensee must actively promote the invention or rights will be terminated.
- Licensee will enforce the patent (usually only exclusive licensees will agree to this provision).
- Inventor is indemnified against claims by third parties resulting from licensee's use of the patents.
- Provision of product liability insurance.
- Provision for licensee's records to be reviewed by inventor.

Consider your relationship with your licensee to be a partnership in which both of you seek to make money on your product and to share that money on a risk versus reward basis. Stay involved with the company that has licensed your invention.

SUCCESS STORY

David Jisa, Game Inventor of Roll-It Tic-Tac-Toe

Ever invented a game and wondered how to get it out there? David Jisa did just that. As a teacher, Jisa used games in his classroom for years to help motivate students to learn. Some of the games, such as Roll-It Tic-Tac-Toe, were his own inventions.

"In creating any game," Jisa says, "I try to involve different methods of learning. For Roll-It Tic-Tac-Toe, I was able to incorporate mathematical logical learning through placement of numbers on the game board and interpersonal learning by playing with others."

Because his students liked Roll-It Tic-Tac-Toe, he thought others might have fun playing it, too. Jisa advises potential game inventors, "Develop games that are easy to understand, can be played by a broad age group, and won't cost much to manufacture."

Looking for a licensee, Jisa researched Patch Products, a leading manufacturer and marketer of family entertainment products. On their Web site, www.patchproducts.com, he found step-by-step instructions on how to submit proposals for new game ideas. He followed the steps. Patch liked his game and licensed it.

Toy and Game Inventing

Toy and game inventing operates somewhat differently from the rest of the inventing industry. Toys and games are almost never patented as the industry is very fast moving and inventors would be left behind the trends if they waited to receive patents for their products. Inventors should, however, conduct a patent and market search to make sure the product does not already exist. Spotlight on Games (www.spotlightongames.com) has details about hundreds of games in existence.

Ask the Inventors (www.asktheinventors.com) offers licensing referral links to toy agents in the U.S. The Web site http://zachsarette.com/inventingtoys.html is a good resource in the field of toy inventing and gives detailed information about utility patents, design patents, trademarks, and copyright for toy inventors.

The Toy Industry Association (www.toyassociation.org) says that an independent inventor's best chances may lie with contacting small- and medium-sized manufacturers directly. Be certain the manufacturer you contact and your product are compatible. You can determine a company's product line through your toy store research, or obtain a list of toy manufacturers that includes a product line description. Call the companies directly to ask if they accept outside ideas, and if so, to whom and where you may address correspondence.

Before a manufacturer asks to see your invention, you will probably be asked to sign what is referred to as a disclosure or idea submission form or agreement. These forms will vary in content; their primary purpose is to protect both you and the manufacturers, as they establish exactly what you have revealed to them, and at the same time release them from certain liabilities regarding what has been disclosed.

If a manufacturer wants to buy your invention, a royalty agreement is usually made between both parties. Royalty payments usually range from 2 to 10 percent of the item's gross sales, with 5 percent being the average. Be prepared to receive a lower royalty on a licensed product.

The Toy Industry Association's Web site, www.toyassociation.org, also has comprehensive advice regarding toy fairs and trade shows.

Manufacturing

I always thought the best thing in the world would be to have your own product.
—Bob Merrick, inventor, entrepreneur, and author

SOME INVENTORS CHOOSE THIS ROAD AND FIND FUN AND excitement in the quest for success. They become inspiration to future entrepreneurs. Going into business is a major undertaking and takes time, money, and energy. The problems for an independent inventor who is manufacturing and distributing a product are many.

Some inventors, unable to find a licensee for an unproven product, go into manufacturing their product with the idea of eventually licensing it and perhaps obtaining higher royalties for the proven product. As nothing succeeds like success, this could be a very viable road to take. If you can prove your product is a success even on a small scale, people will start to look at it with interest. Trade shows, catalogs, and the Internet are all ways to introduce your product on a small scale and establish some success. Make sure you document your sales with figures you can easily share.

How can you decide if your product is better suited for licensing or starting a company and going into business? Your chances of success in either field are usually based on the distribution channel's acceptance of new products from a "one product" vendor, and the size and specificity of the market in your related field. Some industries are more difficult to penetrate than others and many retail stores will not buy from a one-product vendor.

Industry fragmentation refers to how few companies control most of the market share in an industry. If four or fewer companies control 60 percent or more of the market share in an industry, the industry is considered consolidated. These companies are extremely difficult to compete with. The U.S. automobile industry, for example, is consolidated and has high barriers to entry.

If your product is simple, physically small, can be made at low cost, has low barriers to entry, and is in a fragmented industry, it may make sense to manufacture your invention or product. What a large manufacturing company may consider a small profit, could actually be a worthwhile profit for an inventor or manufacturer.

SUCCESS STORY

Maurice Kanbar, Highly Successful Inventor, Entrepreneur, and Author

Maurice Kanbar, in his book *Secrets from an Inventor's Notebook*, relates how he took his game, Tangoes, to major game and toy manufacturers who passed on it because their research showed they would only sell a hundred thousand units a year. Kanbar did his math and decided a hundred thousand units at $2 a game unit would give him $200,000 gross a year. Not bad at all! He decided to give it a try.

Tangoes, based on the ancient Chinese tangram puzzle, did terrifically well, which led to Kanbar forming Rex Games in the early 1990s. A couple of years later he and a salesman partner were grossing nearly $2 million a year with games fostering math, language, and reasoning skills in players of all ages. Perhaps they celebrated with SKYY Vodka, the no-hangover alcohol that is another of Kanbar's success stories.

Kanbar says that tenacity is very important. He stuck with his ideas even when people thought he was crazy.

SUBCONTRACTING YOUR MANUFACTURING

You youngsters won't believe this, but there was a time when Americans actually made physical things called "products" right here in America. . . . The making of things was outsourced decades ago to foreign nations such as Asia. Today, we Americans are dimly aware that our TVs, computers, cell phones, underwear, dentures, cartoons, etc., must come from SOMEWHERE, but we have no real clue who is making them, or how. We have enough trouble figuring out how to remove the packaging.

—Dave Barry

If you decide to become a manufacturer, I recommend you subcontract or outsource your manufacturing. Subcontracting and outsourcing the manufacturing of your product will enable you to avoid the big-capital investment of purchasing machinery and paying someone to run it.

Get a quote from manufacturers that make products similar to yours. To locate these manufacturers, go to a large library and check the *Thomas Register* and their products or try the online sources mentioned under the licensing section above. Many manufacturers include advertisements and you can learn about the factories and what they offer. The AT&T directory (on the Internet), also a good source of information, is organized by industry. Trade magazines will also list manufacturers and advertisements.

Another way of finding a good manufacturer is to contact local inventor associations and chambers of commerce.

Advise manufacturers that you are in the process of obtaining a patent or a provisional patent and have the manufacturer sign a confidentiality agreement. (See sample in the Appendix.) Fax requests for quotes on manufacturing. Compare prices and product quality. Ask for costs and quantities. Evaluate the manufacturer's quality and consistency and examine other products they have produced. Ask for references. When you've chosen a manufacturer, ask for free samples or a small run of the product. You can use these to market the product and test its commercial potential with retail merchants.

Overseas Manufacturing

If you need help from a manufacturer in product development, you should choose U.S. manufacturing. Once your product is fully developed you may wish to look into offshore production. The prices you get will probably be much lower than manufacturing in the U.S., but make sure the quotes include things like shipping and marine insurance. Remember you will need to pay customs when importing the product. It is often difficult to find an overseas manufacturer prepared to work in the low volumes you probably initially prefer.

Global representatives, looking for U.S. companies that want their products manufactured overseas, attend many major industry trade shows. Check out trade shows on TSNN, at www.tsnn.com. If you are unable to attend the trade show, call the show sponsor and ask for a show directory. Here you will find contact names and phone numbers for exhibitors and trade councils. You can also track down names of possible overseas manufacturers at embassy trade development councils.

Global trade resource hubs available on the Internet include Alibaba, at www.alibaba.com. The Chinese Internet company introduces manufacturers and buyers and has safe-trading features that help you research seller backgrounds and reputations. FITA Buy and Sell Exchange, at http://fita.worldbid.com, offers showrooms where businesses display their products.

If you do order from an overseas manufacturer, make sure you inspect the first few models off the assembly line to ensure they are what you want. One of the risks in dealing with an overseas manufacturer is that you have to provide an irrevocable letter of credit. This lets your bank transfer money to the manufacturer when the product ships. But what if the product ships before you have a chance to approve the merchandise? Your money may be transferred anyway. One way to minimize this risk is to have the shipment handled by an international freight forwarder. The forwarder can contact you for authorization when the product is ready and you can refuse delivery if the order isn't correct. Forwarders will also assist with a letter of credit, customs, and delivery information. Your state's commerce department should offer import and export assistance.

You can also use a sourcing agent. A sourcing agent with ongoing relationships in other countries can assist you in getting quotes from factories overseas. The agent can also advise you about payments and establishing a relationship with a commercial bank. A good sourcing agent should be able to help you negotiate lower prices and arrange production schedules and shipping dates.

Pricing

Products should be sold for two to four times the manufacturing cost depending on what distribution you use. Be careful to take into account all the expenses involved in manufacturing, warehousing, advertising, and distributing the product. Inventors will usually start off with a low number of sales, and if the profit per sale is too low you could run out of money before your product is profitable. Lowering your price to achieve higher volume could require extra expenses and extra staff. This all has to be taken into account when looking for an optimal price.

One way to determine the value of your invention as a product is to find somewhat similar existing products which are sold to the same kinds of customers in the same kind of way and attempt to work out how much that product is worth. If it is sold through a chain or supermarket, note the normal retail price and knock off 40 percent as the store's margin and 30 percent of what is left as the distributor/wholesaler's margin, to leave 42 percent of the retail price. Margins and distributor's fees vary considerably. If your costs for material, salary, rent, and money set aside for loan repayments come to less than 42 percent of the retail price of the competitor, you are doing well. If you are nowhere near that, you will have to work on reducing costs. (See more at www.inventorresource.co.uk/valueproductevaluation.html.)

Manufacturing to Do List

- A sales plan:
  Get your product photographed for advertising and PR purposes. Develop a plan to locate customers, show products, design a Web page, send flyers, and take orders. Hire sales reps at 10 to 15 percent commission to sell to distributors and retailers.

- A distribution plan:
  Decide how you will store inventory and where. Decide how you will deliver to customers.

- A finance plan:
  Do cost projections of all anticipated expenses. How much sales volume do you need to make a profit?

- A marketing plan:
  Write a detailed marketing plan.

- A publicity plan:
  Send out a new-product press release with a photo to the appropriate trade industry magazines.

- A time schedule:
  Prepare a time schedule to get it all done.

Keep in mind that even when your product is made and selling well, you need to keep thinking of ways to make it better and to keep ahead of the competition. Keep thinking of ways to expand your product line, offer customers something new, and generate attention and interest.

From Nature

Get expert advice and employ experts in areas in which you lack expertise. Author Ken Thompson, in his book *Bio Teams*, refers to organizational biomimetics, using an example of how geese fly in a V-formation with the birds rotating in and out of the lead position. This conserves energy and, Thompson theorizes, is also possibly because no single bird has memorized the whole route.

Collective leadership is the norm in the animal world and groups with rotating leaders possess greater initiative and resilience than a group led by a single executive. Follow nature's formula: Great sales representatives and good marketing and manufacturing people will help you make a success of your invention.

TIP

Tip from Bob Merrick, Inventor, Entrepreneur, and Author of *Stand Alone, Inventor!*

Bob Merrick has what he calls the Stand-Alone Inventor's theory of relativity: "A relatively small, uninteresting market to a large company can be a large, interesting market to an individual inventor." Merrick is the very successful proof of this theory. He has successfully developed and marketed his own inventions for more than thirty years. Some of his useful gadgets include a business-card punch that instantly transforms business cards into Rolodex file cards. He has sold more than 8 million of the punches around the world. They sell for about $8 each in office-supply stores.

Merrick doesn't believe in selling a product idea to someone else. However, he is against inventors investing in costly manufacturing equipment and recommends they outsource production of their products. His company, Merrick Industries Inc., did no manufacturing in-house, rather it outsourced all its production.

To do this, Merrick faxed requests for quotes to appropriate factories and asked for costs on quantities. He selected the best factory by comparing costs, capabilities, location, etc., using the manufacturer's quoted costs as the basis for creating wholesale and retail price lists. With these, the profit margin has to be big enough to allow for product promotional costs, packaging costs, and adequate profit. The end-user price should be about five to fifteen times your cost to manufacture, depending on the category of your product and other factors. He advises developing products that offer repeat sales and products with low manufacturing costs.

Merrick suggests negotiating with your manufacturer to produce a minimum quantity, say, a few hundred or a few thousand units, for sampling. With these samples, you can carry out informal market surveys. It helps if your invention is something small and simple that can easily be transported and demonstrated. His own inventions have always been small and relatively simple, e.g., Crystal-Date watch calendars and Lint Mitt lint removers.

With your samples, and a single-page catalog sheet including your product photo and specifications, carry out a market survey with:

- Appropriate retail merchants. Visit and ask their opinions about the product and how they think it would sell.
- Sales reps working with similar products.
- Pedestrians in shopping malls.

If the product scores well in your surveys, hire local sales reps at 10 to 15 percent commission to sell samples to distributors and retailers. Have the sales reps write advance orders for quantities of your product. With orders to fill, there is less risk in placing that first production order with the manufacturer. Order extra units so you maintain an ample inventory as you line up more sales reps and launch the product across the country. Send out a new-product press release with a photo to the appropriate industry trade magazines to get free publicity. Forward the sales leads to the sales team. And keep promoting and promoting.

Merrick, with a strong background in advertising, is a firm believer in the importance of marketing, and this belief has paid off. His retail products are sold nationally in such stores as Office Depot, Staples, and OfficeMax. He is also a firm believer in helping other inventors and was president for three years of the nonprofit California Inventors Council and a delegate to the White House Conference on Small Business. (See www.bobmerrick.com.)

Kathryn (KK) Gregory, Inventor and Entrepreneur at Age Ten

In 1994, Kathryn (KK) Gregory invented Wristies and became a successful inventor-entrepreneur at ten years of age.

Frustrated by the chilly way snow kept creeping up the sleeve of her jacket, KK had a great idea: Wristies. With her mother's help, she sewed

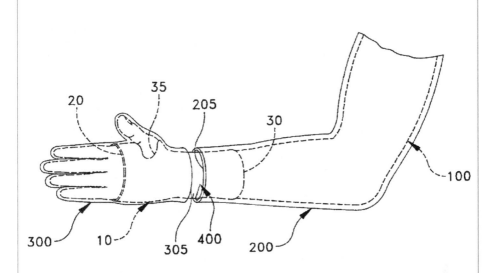

some synthetic fleece into cylinders that would fit snugly over her forearms and hands. She designed the detachable sleeves to extend only as far as the palm of the hand and cut slits for her thumbs to anchor the sleeves in place. KK's first invention didn't work, but she remodeled her product and her next tests were a complete success. She made copies for her Girl Scout troop and knew she had a winning idea when she saw their enthusiastic reactions.

After a lot of research and a consultation with a patent attorney, KK confirmed that her idea was original. She called her invention Wristies and applied for a trademark and patent and founded a company, Wristies Inc.

To promote her invention, KK, from her home office, signed purchase and marketing agreements with the Girl Scouts, Federal Express, and McDonald's, among others. She experienced many challenges along the way. The most difficult one was being taken seriously, as she was so young. But she continued doing what good marketing people advise: Wear your product, use your product, and talk about your product. She lost some friends—lots of kids at school ridiculed and teased her about Wristies—but the Wristies business was growing.

The media was interested in the story of how a young person with a good idea could sell it, start a business, and be an entrepreneur. KK became the youngest person ever to promote a product on QVC, where a six-minute spot earned her $22,000 in sales. She was profiled on the *Today Show,* and in the *Boston Globe* and *New York Times.* Her story provides a great example of the inventing and entrepreneurial process: a novel idea, design of a model, field testing and refinement, market testing, patent search and application, incorporation into a company, promotion, sales, and profits. She won numerous awards, including induction into the Kids Hall of Fame in 1997.

Wristies continue to thrive and there are even heated Wristies. A new product, Sleeves by Wristies, made of Polartec fleece, offers warm soft sleeves in bright colors for nursing home patients with delicate, fragile skin. (See www.wristies.com.)

Marketing

Somebody has to do something, and it's just incredibly pathetic it has to be us.
—Jerry Garcia

Doing business without advertising is like "winking" at a girl in the dark.
You know what you are doing, but no one else does.

—Steward H. Britt, Ad Expert Seward

MANY INVENTORS FEEL THAT MARKETING IS JUST NOT their thing. And many inventions, unmarketed, remain dreams or bits of metal in someone's garage. Others feel that their product is so great that it will sell itself. The reality lies somewhere in between. Licensees and customers will not beat their way to your door. Your invention is a product, and whether you are selling the intellectual property (IP) to licensees or manufacturing and selling the actual product, a basic knowledge of marketing will help you achieve your goal.

Charles Steilen, PhD, is the dean of the College of Business Administration at Hawai'i Pacific University. During twenty-eight years as a marketing expert in Hong Kong, Dr. Steilen helped numerous small- to medium-size businesses in Asia get started and grow. Dr. Steilen helps organizations become market driven and customer focused. His strong belief is that a customer-focused, market-driven business will always have an advantage over the traditionally managed ad hoc type of business.

The strong marketing program that follows is based on a series of articles that comprised Dr. Steilen's series on marketing in *Pacific Business News*.

TIP

Marketing Tips from Dr. Charles Steilen, Marketing Expert

One problem Dr. Steilen sees is that many small- to medium-size businesses believe that because their product and/or service is of such high quality, existing clients will actually promote the product on their behalf. Thus, they build their business around hope that the existing clients will refer new customers to them.

All businesses should be market driven and customer focused, no matter the nature of the business or size of the firm. To achieve this goal requires adopting the viewpoint of the customer. This means basing business decisions on what the customer needs, wants, and even dreams about. And developing products and services that match those needs, wants, and desires.

Providing value is the critical element in becoming a market-driven, customer-focused firm, both for a one-person company or a multinational corporation.

The challenge is for the business owner to become that customer-focused, market-driven firm. A marketing plan, in writing, that focuses on developing marketing strategy for the coming year helps when making critical decisions.

MARKETING PLAN

Determine, define, and prioritize your key target markets. The target market is that group of customers your firm intends to reach. Many businesses have a number of different target markets. Examples include:

- Your existing customer base. You may want to protect this market or, possibly, get them to buy more of your products next year.
- Your competitor's customers. You may want to move some of these customers over to your business.
- Individuals who have never bought into your product category. You may want to get people to become first-time users of this type of product.
- Individuals who may be only one-time buyers, e.g., transient buyers.

Based on your knowledge of your existing customers, your potential customers, market conditions, and your competitors, determine who your key target markets will be in the coming year. Then determine the size of each target market. Then prioritize them.

Develop a sales forecast for each target market. Take only your priority target market for the coming year and determine what you are going to try to achieve in sales. Use your knowledge of market and competitive conditions to arrive at this number.

Formulate your marketing strategy. Strategy is defined as how you deploy your marketing resources. The target market you have considered for this exercise and your sales forecast will now drive the development of your marketing strategy.

MARKETING STRATEGY

A marketing strategy consists of just six strategic weapons. These are:

- Product/service: You may need to add new products, modify an existing product, repackage an existing product, develop a new brand, or even drop a product.
- Customer service support: You may need to consider providing faster service support, adopting a more understanding customer service attitude, or providing additional services.
- Pricing: Price has to reflect the value of what you are offering. You may want to adjust your pricing or consider bundling products.

- Distribution: You may need to expand your channels of distribution or change your location. Are you using the Internet?
- Marketing communications: You may need to undertake some advertising, public relations, or sales promotions.
- Sales force: You may need to adjust the type of approach your sales team is using, perhaps adopting a more advisory or consultative approach rather than a product-focused approach. You may have to increase the amount of networking you do.

Your Key Target Market: Your Existing Customer Base

What has to be done for you to increase sales from your existing customer base? First, undertake an analysis of your existing customers with the objective of building a comprehensive database. This does not just mean collecting names and addresses. It means profiling these customers on what and why they buy, and how often they buy, as well as some indication of their additional needs and preferences. People love to talk about themselves, their families, their jobs, their dreams, and their frustrations. Use this to your advantage. Ask people for their business cards. Keep a record of purchases for each customer. Keep a database.

First, determine what information you need to acquire from your existing customers and how you will acquire this information.

Second, develop a customer-information system that allows you to build up a customer profile.

With this information, determine if there are any unfulfilled needs that exist among these customers. Are these needs unique to individual customers or are there some common needs among the larger group of customers?

You are now in a position to determine where and what the opportunities are for your firm in this market sector. Now set your sales objectives for just your existing customers. Then determine how you will achieve this objective.

To formulate a strategy for this group, ask yourself the following questions. Will you need to:

- Add any product or modify any of your existing products?
- Provide any new added-value customer services?
- Adjust any of your prices to build greater loyalty or create greater consumption?
- Expand your distribution channels?
- Provide more information that introduces them to new products or services, or provides solutions to problems, or just say thanks for their business?
- Change the role of your sales team from telling customers about products to helping customers improve their lives or their businesses?

A marketing strategy directed at your existing customer base is one way to defend yourself from competitors poaching your customers. It can increase revenue and it is a lot less expensive than trying to attract new customers.

Expanding Your Customer Base is Your Firm's Lifeblood

The lifeblood of any business is expanding your customer base. The first step is to identify who your new customers are, how many there are, where they are, from whom they are buying, and whether they have specific needs that your firm can fulfill.

Think creatively. Maybe there's a group of people whose lifestyle or demographic has changed—age, income, marital or family status, etc. Perhaps they don't even know about your firm or its products. It is imperative that you identify, define, and prioritize these customers by individual needs.

- Define who you believe to be your potential group of new customers. What are their characteristics and their needs?
- What is the size of this potential new customer group?
- Why are these individuals or organizations not currently buying from you?
- Why should these individuals or organizations buy from you?
- What are your major competitors doing to attract these customers?
- What can you realistically expect to achieve during the coming year? Do you think you can increase your sales by $5 or $500,000?

The next decision is to determine which of the six marketing weapons you will need to deploy to achieve your objective.

Six Marketing Weapons

- Product and service offering: Based upon knowledge of the needs of this potential-new-customer market segment, is your existing product or service portfolio capable of meeting its needs? If not, what additions or adjustments will you need to make?
- Pricing: Will you need to offer a pricing incentive or add value to acquire or attract a new customer?
- Customer service: Will you have to demonstrate the value of your customer service to acquire a new customer? If so, how would you do it?
- Marketing communications (advertising, public relations, direct mail, Internet): How will you communicate to members of this new market sector the reasons why they should buy from you? What channels of communication can reach this new market? What benefits will you emphasize in your message? What will motivate these people to become your customers?
- Sales force: Your sales team must be prepared to identify these new customers. Some team members may need to be trained to acquire information from these customers and, only then, develop a presentation that is directed to that individual customer.
- Distribution: Depending on the location of your new potential market sector, you may need to consider developing a new distribution location to make buying easier.

Information about potential customers, market conditions, and your competitors is critical in developing a market strategy that allows you to bring in more clients. The information you gather about potential customers will allow you to build a solid foundation on which you can gain a competitive advantage.

Acquiring the relevant information, determining and defining new market segments, setting realistic objectives, and then developing a comprehensive marketing strategy supported by the proper execution of that strategy will always lead to new customers. Keep in mind—there are no shortcuts!

Marketing Quiz

Take this quiz to gauge your marketing efforts.

1. Do you really understand the needs, values, images, and attitudes of the following market sectors:

   Your existing customer base?

   Your competitors' customers?

   Customers who do not currently buy your product but who could become buyers and users?

   Customers who no longer buy from you?

2. Do any of these four market sectors offer you potential for expanding sales?

3. Do you have a written plan?

4. Do you have a sales forecast?

5. If you do have a sales forecast, was it based upon your knowledge of marketing conditions, as well as your knowledge of competitors and customers?

6. Does your company have a proactive customer service support strategy?

7. Do you have an up-to-date customer database?

8. If you have a comprehensive customer database, did you use it in building your marketing strategy for this year?

9. Do you currently use any type of advertising, public relations, sales promotion, or the Internet?

10. If you use any of these marketing communication tools, do they provide specific benefits that customers will experience as a result of buying from you?

11. Do you have any means of generating customer feedback about your business or your competitors' businesses?

12. Do you feel that you need to do something different to expand your businesses?

13. If you do feel that you need to do something, what is it?

Even though you may have a number of "no" answers in this short quiz, this does not mean that your company is not successful. It is intended, however, to identify those areas at which you may want to take a closer look.

Press Releases—Getting Your Invention or New Product Noticed

Good marketing and advertising are vitally important to the success of your invention or new product. In the marketplace, you are competing with big companies that spend billions of dollars every year on print and broadcast ads for new products. You have to promote your invention if you want it to succeed.

Write your own press releases answering the "who, what, when, where, why, and how" questions about your product. Explain exactly what your product is and why you invented it. Focus on what the benefits of your product are to your customers. What will your product do to make their lives easier, or to make them healthier, more beautiful, or happier? Brainstorm with your friends and come up with a catchy slogan.

Send the press releases out to trade magazines, to your local small-business association press, your local neighborhood press, your university newspaper, and "niche" magazines. With the press release, include a catalog sheet with a professional photo of your product. Again list its advantages to consumers. In a separate spot on the page, give the technical facts of your product; its size, weight, and price; ordering information; Web site information; telephone and fax contact numbers; and your e-mail address. (See sample press release in the Appendix.)

Bacon's Magazine Directory lists every U.S. trade magazine under market headings. Before sending your press release, call to find the name of the relevant editor and check articles in the relevant magazine or newspaper to see what the editors are publishing and what they want.

Distribution Channels

Start at your local hardware store, sports store, small specialty stores, or any local store that carries similar products, or products related to your invention. Introduce yourself as a frequent customer, and ask if you can place a few of your products in the store "on consignment." (Make sure you keep careful track of where you place products and start that good bookkeeping system that you are going to need to keep track of sales and who has or hasn't paid you.)

Make frequent visits to the stores to check on your merchandise. Is it displayed well? Do you need to supply more stock? Offer to do demonstrations.

Once you have a record of some successful sales, you can use these to impress other possible vendors in similar stores further from home. Talk to managers at local chain stores like Home Depot. Ask when the district manager is coming to the store and if it would be possible to have a few minutes to talk with him. If the district manager likes your product, ask if he could recommend that it be sold in other stores in the chain on a test basis.

Plan a road trip to see if stores will let you do a demo. Call ahead to say you're coming to their town with an exciting new product and you would like to demo it in their store. Make your demo attractive. Have posters. Free samples. The buzz surrounding your demo draws foot traffic into the store and creates interest in your product.

SUCCESS STORY

VTV Virtual Reality

Software inventors Susan Wyshynski and Eric Gullichsen sold installations of their VTV Mandala system to the Smithsonian Museum, the Washington Children's Museum, and the Library of Congress simply by dropping by and demonstrating their exciting new virtual reality system.

Piling their computers, monitors, and cameras into their van, the two inventors hit the road and headed off to sell their invention, sleeping in the van as they couldn't afford a motel. They pulled up at the Smithsonian Museum and knocked on the curator's door, saying, "Hi. We have a technology we think you'd be interested in. Can we set it up and give you a demonstration?"

The curator was blown away by their demonstration. He saw himself on the computer screen and was able to pick up a paintbrush with his computer-image hand and paint. He could reach out and play the bongo drums. He called other Smithsonian officials to see the product. The timing was serendipitous as the museum was building a new gallery featuring the information age—they asked the inventors to create a customized installation.

On a roll, the inventors headed off with an introduction to the Washington Children's Museum, where they demonstrated and sold VTV games, including a virtual reality game in which a player sees himself on the screen, playing as the goalie.

Active marketing resulted in instant success for their invention.

Niche Markets

Some inventors find success by aiming for a niche market and then modifying the product for another niche market and so on. Niche markets are often easier to break into and the inventor can slowly build a business that reaches all markets eventually.

Sales Reps

Ask local merchants for the names of some sales reps who call with similar products. Contact each sales rep and find out their requirements to pioneer a new product. If you have already established a sales track record, let them know the highlights. Once your product is selling locally, your local sales rep will be able to refer you to manufacturer's reps all over the country.

You can also find sales reps and other useful contacts at trade shows that are held to support distribution of products. When attending a trade show for the industry your product most likely falls into, you will find yourself in the middle of industry distribution channels. Manufacturers, sales people, marketing experts, and the media will all be there, gathered to learn about the latest products, check out the competition, and line up potential distributors.

Search for sales reps by looking for manufacturers' representatives. As a member of the Manufacturers' Agents National Association (MANA), you can search by manufacturers' company names and you will get a list of sales reps that handle various types of products. Some industries have their own sales rep listings, so investigate sales by your specific industry too. Also try the National Association of Wholesalers and its member association list, at www.naw.org/about/assoclist.php.

Baser Door Handles

At seventy-five, retired Sacramento contractor Owen Baser is just getting started as an inventor. Baser has invented the first type of new door handle in more than 200 years, a push-pull design for those with physical difficulties. Baser says, "Getting the idea was the easy part—things start to get hard when the rubber hits the road."

Baser came up with his idea one day when he was trying to get into his laundry room with his hands full and realized it couldn't be done. You needed a free hand to operate a door handle. So he thought about the problem and invented a push-to-enter, pull-to-exit door handle.

As part of his plan, Baser attended a USPTO conference, part of an outreach effort led by John Calvert, administrator of the Patent and Trademark Office Inventors Assistance Program. The office educates inventors through clubs, universities, and schools on how to navigate the complicated patent application process. He used a patent agent in Washington, D.C., to research if the product idea was novel—a more economic method than using a patent attorney.

Baser then used a patent attorney to file for a provisional patent using drawings and a prototype of his design. He eventually acquired two patents with a third pending, an expense of around $30,000.

To promote his product, Baser had a working model of the door handle and showed it at home shows to get response from buyers. He also promoted his product through inventor competitions, winning the 2004 New Product Award in the National Hardware Show and first place and best in show at the Yankee Invention Exposition 2006. He has been interviewed on CNBC and HGTV.

Deciding to manufacture the product himself rather than license it, Baser visited a factory in China and decided China was the way to go. Baser sells his products through his own manufacturing company in stores and at home and garden shows.

"Every day we meet new people and have new challenges," Baser says. "When we learn that we have made someone's quality of life just a little bit better, it is really worth it. We met a combat wounded Navy SEAL

Due to his injuries he was not able to rotate his wrists and could not open a door with a knob. But he could move the tips of his fingers. When he saw and used the Baser Door Handle, you could see the delight in his eyes. He can now use the bathroom in privacy, no need to have his wife open the door for him. He is unable to sign anything, but he can now open doors. Also, service dogs can open doors with the Baser Door Handle." See more at http://baserdoorhandles.com.

Chia Pet

Joe Pedott is a marketing whiz who found an unusual invention, took it to the marketplace, and turned it into a huge success.

In 1977, at the annual housewares show in Chicago, Pedott asked a buyer from a large West Coast drugstore chain about his big holiday sellers. The man told him that something called the Chia Pet always sold out and that he could sell more if they were available. Pedott checked out the product and liked it. The Chia Pet was a little clay figure. When coated with Chia seeds, a Mexican herb, and filled with water, the little figure sprouted a lush coat of green fur in a couple of weeks.

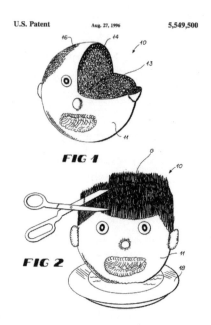

U.S. Patent   Aug. 27, 1996   5,549,500

FIG 1

FIG 2

So Pedott went over to talk to the importer who was importing Chia Pets from Mexico. As he was not making much money, the importer was glad to sell the rights to Pedott. Pedott visited the town in Mexico where the pets were made and discovered the middleman was cheating on prices. He began manufacturing, importing, and advertising the Chia Pet. At an agency meeting, someone pretended to stutter the name and "ch-ch-ch-Chia" became a catchphrase.

Some 500,000 Chia Pets are sold each year. Pedott has found success by seeking unusual yet quality products that can be manufactured reasonably and sold and promoted at a reasonable price. (See www.chia.com.)

Catalogs

Ever sat on an airline and paged through that catalog of fascinating products that you never see in stores? Can you picture your product on one of the glossy pages? Maybe you picture your product in the Brookstone catalog?

There are hundreds of consumer and trade and industrial catalogs that could list and sell your invention. Specialty catalogs can reach people with similar interests throughout the country or even the world. The editors and buyers of these magazines are on the lookout for exciting new products targeted to their specific audience. It is up to you to persuade them that your product will appeal to their readers and that you are an established company who can supply excellent products on time. They usually require large profit margins.

To find the buyers for catalog companies, check the Sunday newspaper supplements and look for advertisements soliciting direct orders for products similar to yours. The *National Directory of Catalogs* lists catalogs and the addresses of their publishers. You can find the directory at large libraries; ask your librarian.

Promo Items

Another way of getting your product noticed is to ask a successful company if they would like to offer it as a promo item, e.g., a discount coupon for your dog bed could be offered with purchase of pet food. Inventor and entrepreneur Bob Merrick turned his Crystal-Date Watch Calendar into a great promo item—companies bought it and gave it as gifts to clients, emblazoned with their name. The navy gave it out with a "Join the Navy" slogan.

Infomercials and Internet Marketing

The Internet provides an opportunity for a small company to compete with big corporations. Search the Internet for marketing sites that will help you promote and sell your product. While your own Web site can be a powerful advertising tool to get your product out there, you need marketing and advertising to draw people to your site and the technology to secure and ship online orders. If you can handle this, your potential profit will be higher than making use of the services of an existing Internet store; however, higher sales volumes are usually possible through established Internet stores because they already have many subscribers and visitors.

Ways of using the Internet include posting a free listing in the Web Yellow Pages, selling through an Internet shopping network such as Amazon Marketplace, using eBay, or purchasing a space at an Internet shopping mall such as Café Press. The top fifteen online shopping destinations listed by Nielsen include four shopping networks with visitors in the millions; check out Shopzilla.com Network, Yahoo! Shopping, Shopping .com Network, and NexTag Network. Many products are also marketed via the billion-dollar television retailer Home Shopping Network. YouTube is another important marketing tool. It's free and has millions of users watching entertaining videos that can also contain informative commercial pitches.

Today's market has changed from a megacorporation-dominated marketplace to a marketplace full of companies run out of the house, either by one person or a team, often husband and wife. Using the opportunities available on the Internet and a virtual office, you can look like a "real" company without renting office space.

WHAT IF YOUR MARKETING ISN'T WORKING?

If your marketing isn't working and not enough customers are buying your product, maybe you are targeting the wrong customers. Would another group of customers value your product more? Do your research in stores, speaking to buyers and customers. Ask sales reps for their advice. You may have more success targeting a special niche in the market instead of targeting everyone. You may need to repackage the product for the new target market. In the cosmetic market, for example, traditionally "female products" like moisturizers have been repackaged successfully for men, or even changed into something similar, but different, like shave cream.

The Home Entrepreneur— Bottom-Feeding

When I think of invention I always think of America. You're always seeing ads:
"Have you got the next big idea?" There seems to be that spirit in America of
inventions and inventors.

—*Simon Cowell*

ALTHOUGH LICENSING IS THE IDEAL FOR A PROLIFIC inventor, it is not always possible to get a licensing or royalty deal. There are some inventions that do not qualify for a utility (or design) patent, but are simply just good ideas for a business.

Having roots in the mail-order business world, I saw the advantage of running a business out of my home, yet still having a professional image. If you can't beat the competition by obtaining a legal monopoly (i.e., a patent), you can still carve out a business, especially if you stick to the principles of bottom-feeding.

DON'T QUIT YOUR DAY JOB SO YOU CAN QUIT YOUR DAY JOB

First of all, never, ever quit your day job. As long as you have income and a front you can show the rest of the world, you can do whatever you

want on the side and there is no pressure to succeed. The reason you don't quit you, day job is so that one day you can quit your day job.

The pressure of trying to make a living while building your invention is too much for most people to take. However, if you look at the numbers, there are 24 hours in a day, minus 6 hours a day to sleep, leaving 18 hours a day to be productive. So that's: 18 hours a day x 7 days a week = 126 hours per week that are available for work. Assuming you have a forty-hour-a-week job, that leaves eighty-six hours per week of available time—that is theoretically enough for two more forty-hour-per-week jobs! This is enough time to develop your invention or business on the side.

By working out of your home, you follow one of the main tenets of a bottom-feeding hardcore inventor: low to no overhead! If you gross $50,000 per year from your home or Internet business, your profit margin is huge: If you have no overhead, you convert most or all of your gross proceeds into net proceeds!

Depending on the type of invention you undertake, you should be able to subcontract out the assistance you need (e.g., lawyers, engineers, and marketing), though you should keep your hired help to a minimum and only when it is required (patent lawyer for your patents or corporate lawyer for your agreements). My father is a perfect example of a person with a great idea, and even though it was not patentable, he stuck to the no-overhead principle.

SUCCESS STORY

David Yonover

My father, David Yonover, is an innovative guy. Many years ago, while he was working for a swimwear manufacturer, he came up with an idea of using stamped metal ID plates (then issued at nightclubs) for social security identification, and invented a customized stamping machine that made the metal-plated social security cards and pet tags. My father thought this was his chance to replace his day job, and he started a business on the side in our garage.

Having a friend who published a magazine in his spare time, my father asked his friend to publish an ad for the social security ID plates. The main pitch lines were: "go into business for yourself" and "earn extra income." It was a dual pitch—buy the metal ID plates for your own personal use or sell the metal ID plates to other people and make a commission. Customers could also order their own machines and start their own home businesses. There were 165 different styles of plates. My father's publisher friend looked at the ad and said, "Dave, you're a good friend, but this idea stinks!" Nevertheless, he humored my father and ran it.

For days after the magazine hit the stands, my father checked his mailbox for replies. Nothing. He thought maybe his friend was right and the idea did stink (at least he still had his day job)!

Finally one day he decided to open the mailbox and just dust it off in there. To his amazement, he saw one card. It was a funny-looking postcard so he brought it to the postal clerk. The clerk went to the back and came back with a giant sack of letters. My father was flabbergasted. And when he started opening the letters, things got even better. Each envelope carried cash payments! He never looked back. That little idea/ad turned into a forty-plus-year business, Perma Products, that put three kids through college. My father is in his mid-eighties and the business is still operational today, with his most popular product being metal pet tags.

The key to my father's business was not patenting his idea, but executing it with a low overhead and marketing the product to the world without a "retail" presence. Forty years ago, he installed an office phone in his bathroom (along with the house phone) and was able to conduct business from his throne, even using a pseudonym, and close deals prior to flushing! That is true success!

THE HOME OFFICE

After your free-form thinking session, you must eventually return to home base. Having been reared in a home-based office since I was a toddler, it was second nature to create one for myself. In my case, as an inventor and scientist, I had to also formulate a home laboratory (in addition to the office). This is a magical time for home offices as the Internet has enabled a person to work twenty-four hours a day all over the world. The cell phone, along with voice mail and caller ID, is another technological advance that extends your home office to just about anywhere in the world.

Soundproofing is the key to the home office. If you have dogs, children, or a boisterous spouse, make sure you have a "quiet" room—ideally one that is well insulated. I prefer cordless phones, which enable seamless and covert transitions from room to room. I am still working on a remote toilet flusher since there is nothing that blows your cover more than the sound of a flush! I have perfected the stretch, flush, and run technique; however, a discerning ear can still pick up the true environment.

Ideally, you should link all the computers in your home, including your children's, to a network so that you can run to any room to take a call and still be able to access the Internet. I am a big believer in redundancy: Multiple computers make it less painful when one crashes. The other cardinal rule for any type of computer work is to always back up your work—especially if it is something creative.

The easiest way to do this is to use a thumb drive. I try to update what I am writing every fifteen minutes, just in case the computer crashes, so I don't lose some golden idea or phrase. If you really want to be thorough about backups, make hard copies to CDs or external drives and ultimately copy them to other computers or send them to your attorney or friend at a different address (in case of house fire or other calamity).

Once you have incorporated a company, you can of course use your home address under the name of your corporation for all business cards and correspondence. Another trick to sounding more legitimate than you may really be is to fluff up your address to make it sound larger and more corporate. Change Apartment #1B to Suite #1B; the mail carrier will still be able to deliver your mail. The other advantage to a home office is potential tax benefits. Check with your accountant to see what is permissible.

When it comes to purely creative thinking, I find it extremely useful to have an unencumbered mind. This is quite a challenge when you are working a day job and running your dream invention out of your house, and even more so if you have a spouse and children.

I like to make sure my mind has a clean slate to think of the good stuff—the highly creative things or applied problem solving necessary to make significant advancements. My low-tech solution for this starts with notes to myself. My office, desk, laboratory, and home itself are full of small pieces of paper with notes to myself that range from the trivial to the complex. Once I have recorded a thought or requirement, I no longer worry that I will forget to accomplish that task (small or large), but I will free my mind to tackle the big stuff—the problems or challenges that require the intense free-form thinking. It has been my experience that the best free-form thinking happens far from your office or even from other humans. The ocean or a solo adventure in nature in any form does it for me. Try to find an escape that works for you; it may be running, working out, or walking the dog.

PROTECTING YOUR NONPATENTABLE INVENTION

If your invention is a product that is not necessarily patentable, you can protect it in other ways. The first and easiest thing to do is form a corporation. Then you can come up with a name for your product and put the trademark sign next to it once you have searched your state and federal records to see that it is not already taken. The "R" is the registered trademark and that will cost you $1,000 or more if you want to pursue it. (See the "Protect" section for details.)

GETTING YOUR PRODUCT OUT THERE

With the advent of the Internet, there are now many more avenues for hardcore home entrepreneurs to get their products out there. I was brought up in a mail-order business where direct mailing and economical response advertisements were the rule. If you can locate your potential customers and present your product to them (without pissing them off with low-level tactics such as spamming), you can reap the benefit of direct sales without using a distributor, middlemen, or, heaven forbid, a retail store!

For example, if your invention or product helps pregnant mothers, then contact them through the media: magazines they read, Web sites they frequent, or shows they watch. There is a fine line between spamming people via the Internet or even direct mail and contacting people that have agreed to be contacted. Of course with eBay and other internet selling methods, a hardcore inventor can really do well, especially if the product is unique and the inventor has at least a rudimentary branding presence.

Ask Expert Advice

Although as a hardcore inventor you may be tempted to do everything yourself, it may be wise to bring in experts for branding or marketing expertise. A good example here would be the genesis of SEE/RESCUE Corporation, my company. My original attempt at the name was SEE LOCATOR. I employed the services of a professional advertising executive (who happened to be my brother and was willing to help me for free). The important point was that I recognized that I needed help and that although I think I am talented, I couldn't do it by myself.

Learning to let go and let someone else run with your idea is an excellent exercise in discipline and ego checking. It remains one of the keys to my success. I am always willing to let anyone voice their opinion and debate the merits of my inventions at any time. Although this often leads to heated debate or banter, good things come out of it. You have to be able to sit back after the dust settles and reflect on the ideas that were tossed around and carefully choose the ones that make the most sense, regardless of whether you thought of them or not. Avoid the "not invented here" syndrome where you are so self-absorbed that you can't recognize other people's good input just because it isn't yours. This also enables you to more effectively work with your consultants.

In a few brainstorming sessions, my brother came up with the "/RESCUE" to make it SEE/RESCUE and the slogan "You Have to Be Seen to Be Rescued." Brilliant! I could have wrapped my brain around it forever and never come up with that. I had to have an open mind and not be an egomaniac to embrace his new name for my technology.

External Funding

If your day job can't keep up with the costs of your in-home venture, you may have to get outside money. There is a saying that the first level of funding is "friends, families, and fools." I would choose families first, since borrowing money from friends often ends poorly. It's harder to break up the blood bond of family. External money can be used to keep a low-overhead home effort going; however, be careful to whom you become indebted.

The problem with borrowing money is that in most cases you have to give up ownership and even control. It is much cleaner to keep it self-funded to avoid those issues. If a friend lends you money, your relationship will change and they will likely feel compelled to provide input, whether you want it or not.

On a larger scale, if your aspirations are for brick and mortar (not low overhead) you may want to consider the conventional routes. Bank loans (including Small Business Administration federal loans) are good because you don't have to give up ownership, but they are usually not large enough. If you think you have a home-run idea that will take hundreds of thousands of dollars to develop, you will likely end up flirting with angel investors or VC (venture capitalists or "vulture capitalists," as they are referred to mostly by people they turn down). Although the money is good, the price of that money is steep, as they usually take a large chunk of ownership and control. They may even fire you and bring in more experienced management. I have known people that were so excited and happy to get VC funding, only to be devastated when they were relieved of their position in the company. (They still retained some ownership, but talk about loss of control!)

Venture Capitalists and Angel Investors

Curiously, this English language abbreviation OPM for other people's money is onomatopoeic, i.e., quickly pronounced sounds like O-Pi-uM—and, why not? Both opium and other people's money are highly addictive and prone to tempt us to abuse.

—*Louis Brandeis*

BOTH VENTURE CAPITALISTS AND ANGEL INVESTORS invest in entrepreneurial firms and take equity in the business. In general, venture capital firms finance later-stage businesses and invest $2 million or up; angel investors invest in early-stage businesses and make smaller investments. Many start-ups find it necessary to self-finance—bootstrapping—until they reach a point when they can credibly approach angels or venture capital firms.

Venture capital firms come in various sizes from small specialist firms to firms with over a billion dollars in invested capital around the world. Venture capitalists tend to focus on particular industries. When conducting your research for venture capital investors, you should look for venture capitalists that specialize in your industry. Industries that venture capitalists favor at the time of writing include high technology, green technology, biotech, medical instruments, health care, retail, computers, software, networking, and the Internet. Some funds focus on early-stage start-ups and others focus on late-stage companies that need money for expansion. Before you try to contact a venture firm, check out its Web site to see what it focuses on and what investments it has made.

When it comes to financing, take your time. Check your local chamber of commerce or state department of economic development for referrals to venture capital firms. Watch the business pages of the newspaper for venture capital association meetings where you will meet investors, lawyers, and CEOs. The site for The Funded, www.thefunded.com, has a good database of venture firms and invites members to rate the firms. The result is an interesting mix of happy and horror stories.

Be wary of companies that ask outrageous application fees and ask you to divulge specific information about your product. Ask for references from CEOs of companies to whom an investor has committed funds. Protect your intellectual property. Invite legitimate investors to see a demonstration of your product first. They will see what it does, but not how it does it. Pay an attorney for an up-front consultation and refer to him as a part of your team. Eventually you have to trust someone, but do your due diligence first.

Check with the investor group's Web site to find what documents are required. You will need to have an executive summary of your business ready and a business plan prepared. Check out the SBA's Web site, www .bplans.com/samples/sba.cfm, for samples of varied businesses. Be ready to make your pitch at a meeting of the investor organization and have answers prepared to the following questions:

- What is your product and why is it unique?
- What are your product's advantages over the competitors?
- Is there a large market for your product and who are the customers?
- Do you have existing customers?
- What are your plans for further market penetration?
- How much will your customers pay?
- How much profit will you make and when?
- How much money do you need and for what?
- What are your start-up and development costs?

- How much have you invested and how much are you planning on investing personally?
- What is your experience in what you propose? Do you have a strong management team with experience and proven skills?
- Can you demonstrate that the business is likely to grow rapidly in the next three to seven years?
- Is there an exit strategy for the investor that is reachable within five to seven years?
- What is the potential for a strong return on investment?

TIP

From Guy Kawasaki—Venture Capitalist, Entrepreneur, and Author

Guy Kawasaki is founding partner and entrepreneur-in-residence at Garage Technology Ventures and cofounder of Alltop (www.alltop.com) an "online magazine rack" of popular topics on the Web. Previously an Apple Fellow at Apple Computer Inc., Guy is the author of eight books, including *The Art of the Start, Rules for Revolutionaries, How to Drive Your Competition Crazy, Selling the Dream,* and *The Macintosh Way.* Check out his very informative Web site at www.guykawasaki.com.

HOW TO GET THE ATTENTION OF A VENTURE CAPITALIST, BY GUY KAWASAKI

- Get an introduction by a partner-level lawyer. He should work at a firm that does a lot of venture capital financings like my buddies at Montgomery & Hansen. Best case e-mail/voice mail: "This is the most interesting company I've seen in my twenty years of legal work for start-ups." Venture capitalists dream about calls like this—it's the equivalent of a scoring shot that knocks the goalie's water bottle off the top shelf.

- Incidentally, part of the reason why you should pay top dollar and use a well-known corporate-finance attorney instead of Uncle Joe the divorce lawyer (even if he handles venture capitalists' divorces): You're paying for connections, not just expertise.

- Get an introduction by a professor of engineering. Best case e-mail/voice mail: "These students are the smartest ones I've ever had in twenty years of teaching computer science. Larry and Sergeiy would have carried their backpacks for them." Arguably his is even better than the lawyer's call if the school has a history of receiving multi-million-dollar donations from its alumni—if you know what I mean.

- Get an introduction by the founder of a company in the venture capitalist's portfolio. Best case e-mail/voice mail: "My buddies are starting a new company, and I think it's really cool." For this to work, it would help if the person making the call is a successful company in the venture capitalist's portfolio. Also this would be a good time to tap your network in LinkedIn to find acquaintances in the portfolio.

- Here's a power tip regarding getting venture capitalists using LinkedIn. Maybe it's only me, but I hate when a connection of a connection of a connection wants me to take a look at a deal. LinkedIn enables you to just go direct, and that's my advice if you can show success (see below). If you can't show success, the connection of a connection of a connection is useless anyway.

- Show success. Suppose you can't get any of the introductions mentioned above. Then the most compelling e-mail/voice mail that you provide is this: "My buddy and I have been working in our garage, taking no pay, and with MySQL we built a site that is doubling in traffic every month. Right now, we're at 250,000 page views a day after thirty days." You've proven you can make a little bit of money ("none") go far, your architecture looks scalable, and most important, the dogs are already eating the food.

- Another way to show success is to hit it out of the park at Demo or the poor man's Demo we call Launch: Silicon Valley. But this is a game that only a few dozen companies can play in every year. Finally, you can provide links to articles singing your praises, but this only means that you fooled the press, not that the dogs like what you're serving.

- Make sure your company is in the right space. No matter how you get to the venture capitalist, make sure that he is the right one for you. For example, if you have the cure for cancer, contacting a firm's enterprise-software guru isn't the brightest idea, so get on the Web and do your homework.
- Use a short e-mail. The ideal length of your e-mail is three or four paragraphs:
What does your company do?
What problem are you solving?
What's special about your technology, marketing, expertise, or connections?
Who are you?

Here are some things not to do:

- Attach a PowerPoint presentation. Save it for the face-to-face meeting.
- Use the word "patented" more than once. All it takes to file a patent is $1,000. No good venture capitalist believes patents make your company defensible. They just want to learn (once) that here might be something worth patenting.
- Claim that you're in a multibillion-dollar market. Isn't every company in a multibillion-dollor market according to some study? At least every company that's ever pitched a venture capitalist.
- Provide a lofty financial projection. Most projections that I see show how you'll grow faster than Google. Frankly, I wouldn't provide any projection at all. It will be either too low and make your deal uninteresting or too high and make you look delusional.
- Brag about an MBA Most venture capitalists want to invest in hard-core engineers at the start. The MBAs can come later, so focus on engineering or avoid the subject completely.
- Try to create the illusion of scarcity. Many entrepreneurs claim that "Sequoia is interested." If Sequoia is interested, you should take its money. If it isn't, then the venture capitalist won't be either. Either way, don't even think of blowing this smoke.

ANGEL INVESTORS

An angel investor is a wealthy individual who invests his or her own money in start-up companies in exchange for an equity share of the business. ACEF (Angel Capital Education Foundation) recommends that inventors and entrepreneurs look for investors who are accredited investors who meet the requirements of the Securities and Exchange Commission. (See www.angelcapitaleducation.org.)

Angel investors are experienced entrepreneurs who can often provide valuable management advice and important contacts. Often former entrepreneurs, they make investments in order to gain a return on their money and also to participate in the entrepreneurial scene and give something back to the community.

Because angel investments bear high risk, they require a high return on investment. Professional angel investor groups look for investments that have the potential to return at least or more times ten their original investment within five years. As angel investors want to be compensated for risk, the earlier they invest, the higher the risk, and the more equity stake they will require in your business.

The investing process starts with a pre-money valuation of the business or venture. If few milestones have been met, the pre-money valuation will probably be low, and the angel investor will expect equity in proportion to the funds he is investing. If you have a prototype, a patent, and customers, and your company has been around for a while, you will be in a better position to negotiate favorable terms with an angel. In considering the ownership stakes, you must weigh the risk/reward of taking on the angel investor against the dilution of control and ownership in your company and whether you want to own a smaller percentage of a larger company.

Because angel investors are more likely to invest in someone recommended by people they know and trust, it is important to network. Possible sources of a referral are: other entrepreneurs who are backed by angel investors or venture capitalists, attorneys who specialize in equity investment bankers, accountants, and business counselors.

The Art of Raising Angel Capital—Guy Kawasaki

Make no mistake about it: There is an art to raising angel capital. Raising angel capital is not harder or easier than raising institutional venture capital—it's simply different. Here's how to do it:

- Make sure they are "accredited" investors. "Accredited" is legalese for "rich enough to never get back a penny." You can get into a boatload of trouble for selling stock to the proverbial "little old lady in Florida," so don't do it. And get a good corporate-finance attorney (as opposed to your aunt the divorce lawyer) to advise you about the process of seeking investments.

- Make sure they're sophisticated investors. I'm a masochist for hate e-mail, but I'll tell you anyway: The least desirable angel investor is a rich doctor or dentist—unless you're a life-sciences entrepreneur. They will drive you crazy because they read how Ram Shriram made gazillions of dollars as an early investor in Google, and now, six months later, they want to know when you're going public too. Sophisticated angel investors have knowledge and expertise in your industry—they will have "been there and done that." Sure, you want their money, but you also want their expertise. Be warned: If you want to raise venture capital in later rounds, it's going to be much harder if you show up with a long list of unsophisticated investors.

- Don't underestimate them. If I had a nickel for every time an entrepreneur told me that he was going to raise angel capital because it was easier than raising venture capital, I wouldn't have to run ads on my blog. Do everything on the venture capitalist wish list because the days of angel investors as "easy marks" are gone forever—if this was ever true. You can have an "early stage" company but not a "dumb ass" company, and angels care as much about liquidity as venture capitalists—maybe even more since they're investing their personal, after-tax money. Angels do not consider investments to be "charitable contributions"—well, no angel whose money you'd want, anyway.

- Understand their motivation. Here's how angel investors differ from venture capitalists: Typically, the angel investors have a double bottom line. They've "made it," so now they want to "pay back" society by helping the next generation of entrepreneurs. Thus, they are often willing to invest in less proven, more risky deals to provide entrepreneurs with the ability to get to the next stage. I know many nice venture capitalists, but I cannot tell you that any of them is motivated by the desire to pay back society.

- Enable them to live vicariously. More on angel motivation: One of the rewards of angel investing is the ability to live vicariously through an entrepreneur's efforts. That is, angels want to relive the thrills of entrepreneurship while avoiding the firing line. Thus, you should frequently seek their guidance because they enjoy helping you. By contrast, most venture capitalists only want to get involved when things are going really well or really poorly.

- Make your story comprehensible to a spouse. The investment committee for many venture capitalists works like this: "You vote for my deal, and I'll vote for yours." That's not how decisions are made by angel investors, because the dual membership of an angel investment committee consists of one person: a spouse. So, if you've got a "client-server open source OPML carrier class enterprise software" product, you must make it comprehensible to the angel's spouse when he or she asks, "What are we going to invest $100,000 into?"

- Sign up people who've heard that angel investors are also motivated by the social aspect of investing with buddies in start-ups run by bright, young people who are changing the world. Even if the other investors are not buddies, investing side by side with well-known angels is quite attractive. If you get one of these guys or gals, you're likely to attract a whole flock of angels too.

- Be nice. More so than venture capitalists, angel investors fall in love with entrepreneurs. Often, the entrepreneurs remind them of their sons or daughters—or fill the position of the sons or daughters they haven't had. Venture capitalists will often invest in a schmuck if the schmuck is a proven moneymaker. If you're seeking angel capital, you're probably not proven, so you can't get away with acting like a schmuck. Therefore, be nice until you're proven—although I hope that when you're proven, you'll also realize that you should be a mensch. (See http://blog.guykawasaki.com.)

ANGEL INVESTORS

Google (Research)

The most famous angel investment in recent years is the $100,000 check that Sun Microsystems cofounder, Andy Bechtolsheim, made out to Google after watching Larry Page and Sergey Brin's demonstration of their search engine software. The check wasn't cashed at first because Google as a legal entity didn't exist. But once the company's incorporation papers were filed, the money helped Page and Brin move out of their dorm rooms. And what would we do without Google now?

FeedDigest

Peter Cooper, founder of FeedDigest, put a PayPal button on his Web site asking users of his online application to make donations. He received more than $5,000 in donations and also attracted the attention of angel investors Kelly Smith and Adrian Hanauer of Curious Office Partners, Seattle. Noting Cooper's online application, which automatically displays news, blog posts, and other content on a single Web page, and the fact that he had thousands of users keeping his project going, Curious Office Partners, an angel investment company that works with software and Web companies, advanced $100,000 in angel money. (See www.money.com.)

Cooper sold Feed Digest in July 2007 to a Russian company called Vicman Software. The service, Cooper says, is basically suited for anyone who wants to use, combine, or republish news feeds (from blogs, news sites, or calendars). The service can mix and republish these feeds into automatically updating JavaScript codes that users can add to their sites, or as images for social profiles, and a number of other more technical formats. It can also be used by people who want to mix multiple feeds into one or perform queries upon those feeds to extract only stories that match those queries.

Curious Office Partners congratulated Cooper on his successful sale and looks forward to working with him again in the future. Presently, Cooper is running a small network of blogs and is on the lookout for another larger-than-life-project to sink his teeth into. (See www.rubyinside.com.)

ROYALTY FINANCING

With royalty financing, an investor lends you a sum of money to help launch your product in return for a royalty on every unit of product that you sell. If your product fails, the investor loses, not you. However, if the product succeeds, the longer you pay the investor after the break-even point, the greater the cost for the money.

When negotiating for royalty financing, you may want to put a cap on the total amount of money you are obligated to pay your lender. Whether or not they agree probably will depend on the size of the royalty. Be sure to work out the numbers with a financial advisor before signing an agreement, and have the agreement prepared by an attorney.

PARTNERSHIP MONEY

Many inventors look for a silent or active partner willing to put money into the venture under a partnership agreement. Unfortunately it is as hard to find compatible business partners for the long haul as it is to marry and stay married. Make sure the partnership agreement lists the duties of each partner and the financial rewards, and have the agreement professionally prepared.

In picking a partner, it's a good idea to choose someone who can do stuff that you can't. If you have good sales and marketing skills, you could look for a production partner involved in your field. Your local inventors association may be able to steer you to a manufacturer in your field who has experience in working with inventors. If you are technically competent but can't sell, consider adding a distributor or manufacturer's sales rep to your team. How you split the profits is up to negotiation and the relative strengths each party brings to the table.

Critical Crossroads: Dealing with Setbacks

If you think you can, you can. If you think you can't, you're right.

—Henry Ford

If at first you don't succeed, you're running about average.

—M. H. Alderson

We don't like their sound, and guitar music is on its way out.

—Decca Recording Company, rejecting The Beatles in 1962

ON YOUR JOURNEY, YOU WILL COME TO MANY CROSSROADS where you have to make a decision that could affect the whole future of the invention: Should you invest more? Should you license it cheaply to get it out there? Should you quit your day job? (My rule is not until you have actual cash flowing from your invention). Should you pay for a patent or a trademark?

Answers and actions when you encounter these critical crossroads should not be taken lightly. Mull over these decisions long and hard and go with your heart. In many cases, these types of decisions define you as a person. Try not to go into "paralysis by analysis" mode, but think about it enough to become comfortable with your decision. Sometimes the timing isn't right, so it might make more sense to sit on your idea until a better time presents itself.

These are personal and many times soul-searching and life-defining moments. Don't take them lightly, but enjoy the ride, because when you are at this stage, and there are many, you have truly achieved entrepreneur status! Persevere, and don't quit your day job—success will follow!

APPENDIX

USEFUL WEB RESOURCES AND LINKS

www.inventnet.com—Invent Net Community: An informational forum that brings together independent inventors, patenting professionals, marketing specialists, and business experts.

www.uiausa.org—National Inventors Association: Organization for inventors, with a wealth of information and resources.

www.inventorsdigest.com—Inventors Digest: A magazine that has useful articles available on all aspects of inventing.

www.invent.org—National Inventors Hall of Fame: Information for inventors.

www.inventorfraud.com/inventorgroups.htm—Inventor Fraud: Inventor organizations state by state.

www.asktheinventors.com—Ask the Inventors Informative: Web site for inventors.

www.tenonline.org—Zimmer Foundation: Offers help with product development.

www.energytechnet.com—Energy TechNet Inventions: Innovation toolbox for energy technology developers.

www.centerdesignbusiness.org—The Center for Design and Business: Programs and services for designers, inventors, and entrepreneurs bringing new products to market.

www.rjriley.com—R. J. Riley's Web site: Offers inventor resources, general information on patents and advice on scams.

www.younginventorsinternational.com—Young Inventors International: This nonprofit organization provides resources to help bring new ideas and technologies to the market for innovators under the age of thirty-five.

www.sbaonline/sba/gov—Small Business Administration: offers expert advice, assistance, and funding.

www.inventright.com:—Inventor Resources and success stories.

www.inventhelper.org—Minnesota Inventors Congress: Provides year-round information to inventors and hosts an annual invention show (Tel: 800-INVENT-1).

www.tsnn.com: The ultimate directory of trade shows.

http://www.tradeworld.co.uk/serve.php—Tradeworld Exhibition Centre: Information on worldwide trade fairs, expositions and shows.

www.inventionconvention.com—Invention Convention: An online cyber tradeshow and newsletter, forum to meet potential investors, manufacturers, and distributors.

MARKETING AND LICENSING

www.bigideasgroup.net—Big Ideas Group: Connects inventors with innovation-seeking companies.

www.sbaonline.sba.gov/starting/businessplan.html—Small Business Administration: offers thirty-one page outline for writing a business plan.

www.business.gov:—A great one-stop shop for people wanting to start small business ventures.

http://tenonline.org/index.html—Ed Zimmer's Entrepreneur Network: Help for Midwestern inventors and entrepreneurs; offers resources and networking connections.

www.wdfm.com—Web Digest for Marketers: Reviews of marketing sites on the Internet.

www.inventorsdigest.com—Inventors Digest: Interesting and informative Web site with a patents issued column.

www.inventnet.com:—Assistance for inventors and free advertising of inventions for sale.

ENERGY TECHNOLOGY DEVELOPERS

www.sbaonline.sba.gov/hotlist/procure.html—SBA Procurements & Grants.

http://grants.nih.gov/grants/index.cfm—National Institutes of Health Grants & Funding.

www.dodsbir.net/—The Department of Defense (DoD): SBIR and STTR programs fund a billion dollars each year in early-stage R&D projects at small technology companies—projects that serve a DoD need and have commercial application.

www.sba.gov/SBIR/—Small Business Innovation Research Program (SBIR) is a program that encourages small businesses to explore their technological potential and provides the incentive to profit from commercialization.

GOVERNMENT GRANTS

www.nrel.gov/business_opportunities/small_business.html—National Renewable Energy Lab: NREL's focus on small, disadvantaged, women–owned, HUBZone, veteran-owned, and disabled veteran–owned businesses extends far beyond the purchasing and subcontracting dollars, which are an integral part of the high performance standards set forth in all areas of the laboratory.

VENTURE CAPITAL, ANGEL INVESTORS, INVENTION FUNDING AND ADVICE

www.nvca.org/resources/otheruseful.html—National Venture Capital.

www.nvca-e-series.org/scriptcontent/membersites.cfm—National Venture Capital Association members.

www.nasvf.org/os/nasvf.nsf/members.html—National Association of Seed & Venture Funds Members.

www.angelcapitaleducation.org/dir_resources/directory.aspx—Listing of angel investment groups in the United States and Canada.

www.angelcapitalassociation.org/dir_directory/directory.aspx—Angel organizations directory.

www.vfinance.com—Venture Capital Resource Library: Directory of venture capital firms, investment banks, and related resources and services.

www.ideafinder.com/resource/r-web.htm—The Great Idea Finder funding sources.

www.federalgrantswire.com/inventions-and-innovations.html—Federal Grants: A comprehensive site on available grants and how to apply for them.

www.financenet.gov—Finance Net: Advice and resources.

www.nsf.gov—National Science Foundation: Advice and resources.

www.sbaonline.sba.gov/hotlist/procure.html—SBA Grants hot list.

http://venturecapital.alltop.com/—Alltop.com: Venture capital advice.

http://smallbusiness.alltop.com/ and http://startups.alltop.com/—Guy Kawasaki, founding partner at Garage Technology Ventures, cofounder of Alltop.com, an "online magazine rack" of popular topics on the Web.

www.nvca-e-series.org/scriptcontent/membersites.cfm#n—Venture Capitalists.

ROYALTY RATES WEB SITES

www.knowledgeexpress.com—Knowledge Express: By subscription.

www.royaltysource.com/—RoyaltySource.

www.energy.ca.gov/research/innovations—California Energy Commission: A small grant program supported by the California Energy Commission that provides up to $75,000 for conducting innovative energy research.

SMALL-BUSINESS COUNSELING

www.score.org—Service Corps of Retired Executives: Offers free e-mail counseling and mentoring to small businesses, including pre-start-ups and inventors.

MANUFACTURING (SMALL BUSINESS)

www.thomasregister.com—Manufacturers.

www.harrisinfo.com—Manufacturers (by state).

www.tsnn.com—Trade shows.

www.inpex.com-Inpex—Biggest inventor exhibition in the United States.

www.naw.org/about/assoclist.php—National Association of Wholesalers:
   Member association list.

www.manaonline.org—Sales reps directory—fee for membership access

www.garage.com—For finding angel investors.

www.businessfinance.com—For finding other financing.

www.nolo.com—Legal self-help.

www.sbaonline.sba.gov—Small Business Administration: Offers expert
   advice, funding, and assistance.

INVENTION EVALUATION SERVICES

www.inventorservices.com

www.wini2.com—Wal-Mart's Invention Evaluation Program.

http://academics.uww.edu/business/innovate/product-assesment-
   brochure.htm—Wisconsin Innovation Service Center.

SALES AGENTS

www.manaonline.org—Manufacturers' Agents National Association:
   Membership fee required; association that includes a list of sales
   agents and manufacturers' representatives.

www.vmwininc.com/repsource.html—Site with manufacturers'
   representatives listed by region and product line.

TOYS AND GAMES

www.toy-tma.org—Toy Manufacturers of America: Offers information
   about toy industry, lists shows and events, and addresses safety
   issues.

www.spotlightongames.com: Huge resource for inventors of toys and
    games.
www.kidsinvent.org—Kids Invent toys: Toy-inventing workshops.

FROM THE U.S. PATENT AND TRADEMARK OFFICE

Many inventors attempt to make their own search of the prior patents
and publications before applying for a patent. This may be done in the
Patent Search Room of the USPTO, and in libraries, located throughout
the United States, which have been designated as Patent and Trademark
Depository Libraries. Useful sites:

www.uspto.gov/web/offices/com/iip—Independent Inventor Resources:
    The USPTO will publish, at the request of a patent owner, a notice
    in the official gazette that the patent is available for licensing or sale.
    There is a fee for this service.
www.uspto.gov/web/offices/pac/dapp/pacmain.html—Inventors
    Assistance Center (IAC): Provides patent information and services
    to the public. The IAC is staffed by former supervisory patent
    examiners and experienced primary examiners who answer general
    questions concerning patent examining policy and procedure.
    Phone: 800-786-9199.
www.uspto.gov/web/offices/pac/doc/general/index.html#mark—General
    information regarding patents from the U.S. Patent and Trademark
    Office.
http://appft1.uspto.gov/netahtml/PTO/search-bool.html—US Patent &
    Trademark Office: Patent application database.
http://patents.uspto.gov/—Web patent database.
http://ep.espacenet.com/quickSearch?locale=en_ep—European Patent
    Office: Worldwide search.
www.uspto.gov/web/offices/ac/ido/ptdl/ptdlib_1.html—USPTO
    Libraries.
www.uspto.gov/web/offices/tac/doc/basic—Basic facts about trademarks.

www.uspto.gov/web/offices/com/iip/data.htm#LicensePromotion—
Public forum regarding complaints concerning invention
promoters/promotion firms.

www.uspto.gov/web/offices/dcom/olia/oed/roster/index.html—The
USPTO maintains a directory of registered patent attorneys and
agents.

www.uspto.gov/web/offices/pac/disdo.html—Information about
disclosure documents.

www.freepatentsonline.com—Free Patents Online: This Web site
provides the full text of U.S. patents from number 4,000,000 (circa
1974) to the present. The database is extremely user friendly and
allows visitors to download free PDF files of patent images. It also
has a "Crazy Patents" page with quirky and eccentric listings.

INFORMATIVE SITES

www.ideafinder.com/articles/thelists/index.htm

www.ideafinder.com/resource/r-web.htm

www.si.edu/lemelson—Lemelson Center for the Study of Invention
& Innovation: At Smithsonian Institute; addresses the history of
inventing.

www.invent.org—National Inventors Hall of Fame: Biographical data,
links to invention sites.

http://web.mit.edu/invent—MIT's Invention Dimension: Highlights
different American inventors; offers extensive listing of invention
links.

www.invention-center.com:—Invention Center.

INVENTION EVALUATION

www.uiausa.org—UIA Innovation Assessment Program: A giant first step
in bringing a new product into the marketplace. Offers evaluation
program and resources.

www.innovationcentre.ca—Canadian Innovation Center's Inventors' Assistance Program: Offers evaluation and commercialization services, information, and resources.

www1.eere.energy.gov/inventions—U.S. Department of Energy's Inventions & Innovations Program: Offers financial and technical support (grants) and business training for development of energy-saving inventions.

http://academics.uww.edu/business/innovate/default.htm—Wisconsin Innovation Service Center: Offers technical and market-assessment services for new products.

www.business.wsu.edu/iac—Washington Innovation Assessment Center.

BEWARE OF SCAMS

www.ftc.gov—Federal Trade Commission: Information on dynamics of invention marketing scams and reporting procedures.

www.rjriley.com—Ronald J. Riley's site: Offers listing of known and alleged fraudulent invention-promotion companies by name.

www.inventorfraud.com—Michael Neustel's National Inventor Fraud Center: Information about fraudulent companies and listing of "good guys" inventor resources, including groups, shows, and Web sites.

FIND A SCAM

www.ftc.gov/search: To check if a company has been investigated and/ or fined by the Federal Trade Commission (FTC), enter the word "Invention" in the FTC's Web site search engine.

INSPIRING WEB RESOURCES

www.ideafinder.com/articles/thelists/index.htm
www.ideafinder.com/resource/r-web.htm
www.ideafinder.com/resource/web/rpo-history.htm—History General
www.ideafinder.com/resource/web/rpo-entertainment.htm—History Entertainment

www.ideafinder.com/resource/web/rpo-inventors.htm—History Inventors.

www.ideafinder.com/resource/web/rpo-timeline.htm—History Timeline.

www.ideafinder.com/resource/web/rpo-hallfame.htm—Hall of Fame.

www.invent.org—National Inventors Hall of Fame.

www.ideafinder.com/resource/web/rpo-business.htm—Entrepreneur.

www.ideafinder.com/resource/web/rpo-experiments.htm—Experiments.

www.ideafinder.com/resource/web/rpo-government.htm—U.S. Government.

www.ideafinder.com/resource/web/rpo-magazines.htm—Magazines and eZines.

www.ideafinder.com/resource/web/rpo-patsearch.htm—Patent and Trademark.

www.ideafinder.com/guest/madlist/amd-first.htm—For Inspiration and Recognition of Science and Technology (FIRST): Inspires in young people, their schools, and communities an appreciation of science and technology.

www.ideafinder.com/resource/books/featured_books/raa149.htm—The History of Science and Technology.

http://web.mit.edu/invent—Lemelson-MIT Program: Learn about inventing, inventions, and information on programs and grants offered by the Lemelson-MIT Foundation.

http://web.mit.edu/invent—MIT's Invention Dimension: Highlights different American inventors.

www.invention-center.com—Invention Center Home Page.

www.nesta.org.uk/withflash.html—National Endowment for Science, Technology and the Arts: Funded by a £200 million grant from the U.K. National Lottery.

www.thenameinspector.com—Fun and informative Web site about naming a product and more.

INTERNATIONAL ORGANIZATIONS

www.invention-ifia.ch—International Federation of Inventors Associations (IFIA): A window on the global community of inventors! A great resource for finding international connections and getting a worldwide perspective on invention and creativity.

www.wipo.int/about-wipo/en/index.html—The World Intellectual Property Organization (WIPO): An international organization dedicated to promoting the use and protection of works of the human spirit.

MARKETING AND LICENSING

www.mondotimes.com—Media ownership, Web sites, and contact information. With a free basic membership, you can create a personal media list to store ten favorite media outlets.

www.business.gov—Advice for people wanting to start small-business ventures.

www.uiausa.org—United Inventors' Association.

http://inventors.about.com/od/licensingmarketing/Licensing_Inventions: Comprehensive site with lots of resources.

www.marketlaunchers.com—Resources for inventors, Web site hosting, and listing of new products.

http://tenonline.org/index.html—Ed Zimmer's Entrepreneur Network: Help for Midwestern inventors and entrepreneurs; offers resources and networking connections.

www.usa-canada.les.org—Licensing Executives Society: Offers booklet about basics of licensing and directory of licensing professionals.

www.usa-canada.les.org/consultants/directory.asp—Licensing Consultants & Brokers Directory: Over 150 LES members are included in this new directory.

www.energytechnet.com—Energy TechNet: Inventions and Innovation Toolbox for Energy Technology Developers—A comprehensive, searchable database of national, regional, and state resources for energy technology developers; essential steps in the process of developing and commercializing innovative technologies; hundreds of links to useful sites.

www.marketlaunchers.com—Market Launchers: An invention database for buying or selling products, including an online inventors newsletter.

http://web.mit.edu/cipd—MIT Center for Innovation in Product Development: An MIT program that links academic knowledge with industry partners to research and develop products.

INVENTOR COMPETITIONS

www.invent.org/collegiate/index.html—The Collegiate Inventors Competition: International competition awarding cash prizes to innovative inventions and innovations and research by college students.

www.gwiin.info—Global Women Inventors and Innovators Network: Exceptional inventions and innovations by women; annual conference.

www.mit100K.org—MIT $100K Entrepreneurship Competition: An MIT competition that awards $100,000 to outstanding teams of student entrepreneurs who submit business plans for promising new ventures.

www.staples.com/iq—Staples Invention Quest: Looks for easy new office-product ideas. The contest awards a $25,000 grand prize and the chance to have your product developed and sold in Staples stores.

www.mypetidea.com—Pet Product Inventions: The chance to have that pet product created, and the chance to win $25,000!

www.whirlpool.com/moms—Mother of Invention Grant: Whirlpool Mother of Invention Grant recognizes the ingenuity and business sense of mothers from across the nation. Over $40,000 in grant money is awarded each year!

www.bigideagroup.net—Big Idea Group: Join one of the Big Idea Insight Clubs to help define a new product category and earn prize money and product rewards. Submit your new invention ideas for a chance to have them licensed and developed.

www.createthefuturecontest.com—Create The Future/Design Contest: SolidWorks has teamed up with NASA to demonstrate design and engineering skills.

www.live-edge.com/info—$100,000 Global Environmental Design Challenge: Environmental design challenge with $100,000 prize package!

www.hammacher.com/sfi/sfimain.asp—Hammacher Schlemmer Search for Invention: A contest in four categories: recreation, personal care, electronics, and home and garden. For inventors with a utility patent.

www.nationalhardwareshow.com—The National Hardware Show: Has a number of awards and publicity for new products.

www.invent.org/challenge—Modern Marvels Invent Now Challenge: Named for the History Channel's series; taps a panel of famed inventors, technologists, and industry experts to determine the top twenty-five inventions submitted from nearly 2,500.

YOUNG-INVENTOR CONTESTS

www.aas-world.org-AAS—The Young Inventors' Program Meant To Invent: Program by the Academy of Applied Science offers all students the opportunity for expression and creativity as they develop and practice high-order thinking skills.

www.youthdevelopment.coca-cola.com/home.html—Coca-Cola Youth Partnership: The mission—to empower and inspire young people to realize their potential and their dreams.

www.invent.org/collegiate—Collegiate Inventors Competition: New technology is emerging from colleges and universities across the United States and around the world.

www.nsta.org/programs/craftsman—The Craftsman/NSTA Young Inventors Awards Program: Challenges students to use creativity and imagination along with science, technology, and mechanical ability to invent or modify a tool.

http://school.discovery.com/sciencefaircentral/dysc/index.html— Discovery Channel Young Scientist Challenge: Every year, more than 60,000 children enter science projects in a Science Service affiliated fairs. Six thousand middle school entrants are nominated by fair directors to enter their projects in the Discovery Channel Young Scientist Challenge—the only competition of its kind for students in grades five through eight.

www.dyson.com/designaward—Eye for Why: Student design competition dedicated to encouraging students to think differently by conquering everyday problems with the design of functional products.

www.exploravision.org—The Toshiba/NSTA ExploraVision Awards: A competition for students of all interest, skill, and ability levels in grades K–12 to encourage students to combine their imaginations with science to create future technology.

www.usfirst.org—The FIRST LEGO League (FLL)—Considered the "little league" of the FIRST Robotics Competition; extends the FIRST concept of inspiring children aged 9 through 14 to use science and technology.

www.usfirst.org—The FIRST Robotics Competition: An exciting worldwide competition that teams professionals and young people to solve an engineering design problem. The competitions are high-tech spectator sporting events.

www.futurecity.org—Future City: The National Engineers Week Future City Competition invites middle school students nationwide to create cities of tomorrow.

www.fpsp.org-Future Problem Solving Project—Offers new and exciting learning paths. Students experience the excitement of creative thinking and the thrill of solving difficult problems. Receptive to the needs of all students, FPSP offers competitive and non-competitive programs.

www.kidinventorchallenge.com—Kid Inventor Challenge: Nothing motivates kids like a chance to have fun, be creative, and do something original, like invent a toy. The Kid Inventor Challenge is a real idea sparker, and invites kids in grades 1 through 6 to experience inventing their own toy creations.

www.rocketcontest.org—Team America Aerospace Challenge: Model rocket competition. The AIA and the National Association of Rocketry (NAR) are cosponsoring a model rocket competition for high school students: design, build, and fly a multistage model rocket carrying two raw eggs and an electronic altimeter to exactly 1,500 feet, returning both eggs intact. The top five teams share in a total prize pool of approximately $50,000 in savings bonds, and the sponsoring schools share in approximately $9,000 in cash awards.

www.odysseyofthemind.com—Odyssey of the Mind: An international educational program that provides creative problem-solving opportunities for students from kindergarten through college. Thousands of teams from throughout the world participate in the program.

www.rubegoldberg.com/html/contest.html—The Rube Goldberg Machine Contest: This contest brings the ideas of Pulitzer Prize–winning artist Rube Goldberg's "Invention" cartoons to life.

www.sciserve.org/isef—The Intel International Science and Engineering Fair (ISEF): The world's only science-project competition exclusively for students in the ninth through twelfth grades. Students compete for over $2 million in scholarships, tuition grants, scientific equipment, and scientific trips.

www.siemens-foundation.org/competition—The Siemens Westinghouse Competition in Math, Science, & Technology: Fosters individual growth for high school students who are willing to challenge themselves through science research. Administered by the College Board and funded by the Siemens Foundation.

www.ideafinder.com—For more contests.

Appendix 2

RECOMMENDED READING

Debelak, Don. *Bringing Your Product to Market in Less than a Year.* Hoboken: John Wiley and Sons, 2005.

DeMatteis, Bob. *From Patents to Profit.* Garden City Park: Square One Publishers, 2005.

Gelb, Michael J. and Sarah Miller Caldicott. *Innovate Like Edison.* New York: Dutton Adult, 2007.

Kawakami Kenji. *The Big Bento Box of Unuseless Japanese Inventions.* London: HarperCollins Publishers, 2005.

Kanbar, Maurice. *Secrets from an Inventor's Notebook.* San Francisco: Council Oak Books, 2001.

Michalko, Michael. *Thinkertoys, A Handbook of Creative Thinking Techniques.* Berkeley: Ten Speed Press, 1991

Merrick, Robert G. *Stand Alone, Inventor! And Make Money With Your New Product Ideas.* Lee Publishing, 1997.

Pressman, David. *Patent It Yourself.* Berkeley: Nolo, 2006.

Thompson, Ken. *Bio Teams: High Performance Teams Based on Nature's Most Successful Designs.* Tampa: Meghan-Kiffer Press, 2008.

RECOMMENDED MAGAZINES

Popular Science
Wired

SAMPLE DOCUMENTS

The sample documents that follow are given by way of illustration only and not as legal advice. For your own legal agreements, be sure to consult an attorney.

SAMPLE PRESS RELEASE

The New Retro-Reflective RescueStreamer is now available to the international boating community.

Military-approved RescueStreamer technology is now available with Retro-reflective capability for 24-hour-per-day visual-distress signaling (see attached photos).

Persons lost at sea can now signal their rescuers during day or night hours with the only "continuous" signaling device that is completely "passive" (no batteries, chemicals, or electronics)!

Additional information is included below and at www.RescueStreamer.com.

BACKGROUND

The inexpensive and military-approved RescueStreamer® technology will save lives because it's superior to other existing visual signaling devices, i.e., flares, smoke, and dye markers.

The RescueStreamer® emergency distress signals locate persons lost at sea or on land by providing a continuous signal—without the need for batteries, chemicals, or electronics!

In U.S. Navy certification tests, RescueStreamer® was visible from 1.5 miles away at an altitude of 1,500 feet! U.S. Navy J & A/QR document described RescueStreamer® as "... the only 'passive' and 'indefinite' signaling device that will likely increase the survivor's probability of detection and rescue."

In addition to saving lives, the inexpensive RescueStreamer® technology will save substantial money by:

1. Reducing the annual inspection and replacement costs for expiration-dependent flares, sea dyes, smoke signals, and batteries;
2. Reducing search and rescue operations costs by locating persons quickly and efficiently;
3. Reducing replacement costs associated with training exercises because the RescueStreamer® is reusable.

(The lack of batteries and caustic, flammable chemicals also means safer handling and storage too.)

In addition to the Navy and NAVAIR, all other branches of the U.S. military (including the U.S. Coast Guard) have now approved RescueStreamer® streamers for use by all personnel.

For your reference, please see our Web site (www.SeeRescue.com) for information on our new compact "holster" unit which easily stows on PFD', overboard bags, or on your belt, and provides twenty-five feet of continuous visibility (the unit weighs only five ounces!). Larger units are available for life rafts and boats.

Prices start at US $39.95 for the standard unit and US $69.95 for the new Retro-reflective units.

Advantages of RescueStreamer® over Sea Dye Marker:

1. Works continuously.
2. Works on water or land.
3. Works at night (even better with Retro-reflectors).
4. Discretionary (can be stowed when enemy is around and re-deployed).
5. Does not spoil (Sea Dye can harden and sink).
6. Floats in both stowed and deployed configurations.
7. Attaches to person or life raft (does not drift away).
8. Reusable for training purposes (economical).
9. Works in "rough" as well as "calm" seas.
10. Same basic size and weight as Sea Marker Dye.
11. Environmentally friendly—no pollutants released into water.
12. Marks the person's exact location.

SAMPLE PITCH LETTER

(Dr. Rob Yonover for the RescueStreamer®)

Background

There exists a significant opportunity to save lives in military commands and maximize their operational funding.

Best of all, the means to achieve these important objectives have already been Navy and NAVAIR tested/approved and are available through the U.S. DOD supply system (i.e., NSNs are already in place).

The inexpensive and military-approved SEE/RESCUE® technology will save lives because it's superior to other existing visual-signaling devices, i.e., flares, smoke, and dye markers.

The SEE/RESCUE® high-visibility distress streamers locate persons lost at sea or on land by providing a continuous signal—without the need for batteries, chemicals, or Electronics!

In U.S. Navy certification tests, SEE/RESCUE® was visible from 1.5 miles away at an altitude of 1,500 feet! U.S. Navy J&A/QR document (available upon request) described SEE/RESCUE® as ". . . the only 'passive' and 'indefinite' signaling device that will likely increase the survivor's probability of detection and rescue."

In addition to saving lives, the inexpensive SEE/RESCUE® technology will save commands substantial money by:

1. Reducing the annual inspection and replacement costs for expiration-dependent flares, sea dyes, smoke signals, and batteries.
2. Reducing search and rescue operations costs by locating persons quickly and efficiently.
3. Reducing replacement costs associated with training exercises because the SEE/RESCUE® streamer is reusable.

Therefore, operating budgets will go even farther!

(The lack of batteries and caustic, flammable chemicals also means safer handling and storage too.)

In addition to the Navy and NAVAIR, all other branches of the U.S. military (including the U.S. Coast Guard) have now approved SEE/RESCUE® streamers for use by all personnel.

As a tactical device, SEE/RESCUE® provides a reusable, safer signal for troops to use under hostile circumstances. Unlike uncontrollable smoke signals or flares, which everyone can see, the SEE/RESCUE® streamer is difficult for enemy ground troops to detect when deployed, while remaining highly visible to combat air search and rescue forces.

New IR/retro-reflector option makes SEE/RESCUE® functional twenty-four hours a day.

Vice Admiral USN (retired) Paul Ilg, a Naval Aviator, spent several days on the ground in North Vietnam being hunted by enemy troops before being rescued. After seeing the SEE/RESCUE® demonstrated, he stated, "I probably would have been picked up in half the time. SEE/RESCUE® could have been deployed when I was not under enemy surveillance and re-stowed when the enemy was close."

For your reference, please see our Web site (www.SeeRescue.com) for information on our new camouflage POCKET/RESCUE unit which easily stows in flight jackets, seat kits, life jackets, or life rafts and provides twenty-five feet of continuous visibility.

Save lives and money through the deployment of SEE/RESCUE® streamers!

Sample Mutual Nondisclosure Agreement

(Sample by Dr. R. Yonover)

This Mutual Nondisclosure Agreement ("Agreement") is made and entered into as of June 6, 2008 ("Effective Date") by and between ABC Corporation, a _____ corporation, having a principal place of business at _____ and XYZ Corporation, a _____ corporation, having a principal place of business at _____

ABC and XYZ desire to engage in nonexclusive discussions in connection with the possible establishment of a business relationship between them, and to exchange, for the sole purpose of evaluating the possibility of such a relationship, certain confidential information and documents. The parties acknowledge and agree that the sole and exclusive reason for their exchanging confidential information is to facilitate their discussions of a possible mutually beneficial relationship, and the use or disclosure of any of such shared confidential information is wrongful and in violation of the terms of this Agreement.

To further specifically protect their confidential information, the parties hereby agree as follows:

Definition of Confidential Information

For purposes of this Agreement, the term "Confidential Information" shall mean any and all trade secrets, confidential knowledge, data, or any other proprietary information pertaining to the business of either part. By way of illustration but not limitation, "Confidential Information" includes, in whatever form, format, or media, and however disclosed or learned: (a) inventions, ideas, improvements, discoveries, trade secrets, processes, data, programs, knowledge, technology, drawings, hardware configuration information, know-how, designs, techniques, specifications, formulas, test data, domain names, computer code, other works of authorship and designs whether or not patentable, copyrightable, or otherwise protected by law; and (b) information regarding research, development, experimental work, new and existing products and services, marketing plans and strategies, merchandising and selling, business plans, strategies, forecasts, projections, profits, investments, operations, financings, records, budgets and unpublished financial statements, licenses, prices and costs, suppliers and customers.

Nondisclosure and Nonuse Obligation

Each of the parties, as Recipient, agrees that such Recipient will not, directly or indirectly, on its own or by, for, or through any other person or entity, use, disseminate, or in any way disclose any Confidential Information of the other party, as Discloser, to any person, firm or business, except to the extent necessary for internal evaluations in connection with negotiations, discussions, and consultations with personnel or authorized representatives of such Discloser in furtherance of the purpose stated above, and for any other purpose such Discloser may hereafter author-ize in writing. Furthermore, the existence of any business negotiations, discussions, consultations, or agreements in progress between the parties shall not be released to any public media without written approval of both parties. Each of the parties, as Recipient, agrees that such Recipient shall treat all Confidential Information of the other party, as Discloser, with the same degree of care as such Recipient accords to such Recipient's own Confidential Information, but in no case less than reasonable care. Each of the parties, as Recipient, agrees that such Recipient shall disclose Confidential Information of the other party, as Discloser, only to those of such Recipient's employees who need to know such information, and such Recipient certifies that such Recipient's employees have previously agreed, either as a condition to employment or in order to obtain the Confidential Information of the Discloser, to be bound by terms and conditions substantially similar to those terms and conditions applicable to such Recipient under this Agreement. Each of the parties, as Recipi-ent, shall immediately give notice to the other party, as Discloser, of any unauthorized use or disclosure of Discloser's Confidential Information. Each of the parties, as Recipient, agrees to assist the other party, as Dis-closer, in remedying any such unauthorized use or disclosure of Disclos-er's Confidential Information.

Exclusions from Nondisclosure and Nonuse Obligations.

The obligations under Paragraph 2 ("Nondisclosure and Nonuse Obli-gations") of each of the parties, as Recipient, with respect to any portion of the Confidential Information of the other party, as Discloser, shall not apply to such portion that such Recipient can document: (a) was in the public domain at or subsequent to the time such portion was communi-cated to such Recipient by such Discloser through no fault of such Recipi-ent, (b) was rightfully in such Recipient's possession free of any obligation of confidence at or subsequent to the time such portion was communi-cated to such Recipient by such Discloser, (c) was developed by employees or agents of such Recipient independently of and without reference to any information communicated to such Recipient by such Discloser, or (d) was communicated by such Discloser to an unaffiliated third party free of any obligation of confidence. A disclosure by each of the parties, as Recipient, of Confidential Information of the other party, as Discloser, either (a) in response to a valid order by a court or other governmental body, (b) otherwise required by law, or (c) necessary to establish the rights of either party under this Agreement, shall not be considered to be a breach of this Agreement by such Recipient or a waiver of confidentiality for other purposes; provided, however, such Recipient shall provide prompt prior written notice thereof to such Discloser to enable such Discloser to seek a protective order or otherwise prevent such disclosure.

Ownership and Return of Confidential Information and Other Materials

All Confidential Information of each of the parties, as Discloser, and any Derivatives thereof (as defined below) whether created by such Discloser or by the other party, as Recipient, shall remain the property of Discloser, and no license or other rights to such Discloser's Confidential Information or Derivatives is granted or implied hereby. For purposes of this Agreement, "Derivatives" shall mean: (a) for copyrightable or copyrighted material, any translation, abridgment, revision or other form in which an existing work may be recast, transformed or adapted; (b) for patentable or patented material, any improvement thereon; and (c) for material which is protected by trade secret, any new material derived from such existing trade secret material, including new material which may be protected under copyright, patent, or trade secret laws. All materials (including, without limitation, documents, drawings, models, apparatus, sketches, designs, lists, and all other tangible media of expression) furnished by each of the parties, as Discloser, to the other party, as Recipient, shall remain the property of such Discloser. At such Discloser's request and no later than five (5) days after such request, such Recipient shall promptly destroy or deliver to such Discloser, at such Discloser's option, (a) all materials furnished to such Recipient by such Discloser, (b) all tangible media of expression in such Recipient's possession or control to the extent that such tangible media incorporate any of such Discloser's Confidential Information, and (c) written certification of such Recipient's compliance with such Recipient's obligations under this sentence.

Independent Development

Each of the parties, as Discloser, understands that the other party, as Recipient, may currently or in the future be developing information internally, or receiving information from other parties that may be similar to such Discloser's Confidential Information. Accordingly, nothing in this Agreement will be construed as a representation or inference that such Recipient will not develop products or services, or have products or services developed for such Recipient, that, without violation of this Agreement, compete with the products or systems contemplated by such Discloser's Confidential Information.

Disclosure of Third-Party Information

Neither party shall communicate any information to the other in violation of the proprietary rights of any third party.

No Warranty

All Confidential Information is provided "AS IS" and without any warranty, express, implied, or otherwise, regarding such Confidential Information's accuracy or performance.

No Export

Neither party shall export, directly or indirectly, any technical data acquired from the other party pursuant to this Agreement or any product utilizing any such data to any country for which the U.S. Government or any agency thereof at the time of export requires an export license or other government approval without first obtaining such license or approval.

Term

All Confidential Information of either party disclosed to the other prior to the execution of this Agreement is subject to the terms of this Agreement, and the obligations of the parties hereunder shall continue for a period of seven (7) years from the Effective Date of this Agreement unless terminated pursuant to Paragraph 3 above.

Continuing Obligations

This Agreement shall be binding on and inure to the benefit of the parties hereto and each of their successors and assigns. The terms, provisions, and obligations contained in this Agreement shall survive the termination of any discussions undertaken by the parties.

Notices

Any notices required or permitted by this Agreement shall be in writing and shall be delivered as follows, with notice deemed given as indicated: (a) by personal delivery, when delivered personally; (b) by overnight courier, upon written verification of receipt; (c) by telecopy or facsimile transmission, upon acknowledgment of receipt of electronic transmission; or (d) by certified or registered mail, return receipt requested, upon verification of receipt. Notice shall be sent to the addresses set forth above or to such other address as either party may specify in writing.

Governing Law

This Agreement shall be governed in all respects by the laws of the United States of America and by the laws of the State of Hawaii, as such laws are applied to agreements entered into and to be performed entirely within Hawaii between Hawaii residents. Each of the parties irrevocably consents to the exclusive personal jurisdiction of the federal and state courts located in Hawaii, as applicable, for any matter arising out of or relating to this Agreement, except that in actions seeking to enforce any order or any judgment of such federal or state courts located in Hawaii, such personal jurisdiction shall be nonexclusive.

Severability

If any provision of this Agreement is held by a court of law to be illegal, invalid, or unenforceable, (i) that provision shall be deemed amended to achieve as nearly as possible the same effect as the original provision, and (ii) the legality, validity, and enforceability of the remaining provisions of this Agreement shall not be affected or impaired thereby.

Waiver; Amendment; Modification

No term or provision hereof will be considered waived by either party, and no breach excused by either party, unless such waiver or consent is in writing signed by the party against whom such waiver or consent is asserted. The waiver by either party of, or consent of either party to, a breach of any provision of this Agreement by the other party shall not operate or be construed as a waiver of, consent to, or excuse of any other or subsequent breach by the other party. This Agreement may be amended or modified only by mutual agreement of authorized representatives of the parties in writing.

Injunctive Relief

A breach or threatened breach by either party of any of the promises or agreements contained herein will result in irreparable and continuing damage to the other party for which there will be no fully adequate remedy at law, and such other party shall be entitled to injunctive relief, a decree for specific performance, and such other relief as may be proper (including monetary damages if appropriate) without the need to post a bond or other security, and without the need to prove damages.

Entire Agreement

This Agreement constitutes the entire agreement with respect to the Confidential Information disclosed hereunder and supersedes all prior or contemporaneous oral or written agreements concerning such Confidential Information.

Negotiations

Although the parties are entering into this Agreement to discuss a possible business relationship, there is no obligation on the part of the parties to proceed with negotiations, to enter into a definitive agreement, or to establish a business relationship. Unless and until a definitive agreement is signed by the parties, neither party has any obligation or liability to the other, except in connection with Confidential Information, as set forth in this Agreement. Any discussions undertaken by the parties are nonexclusive, and each of the parties may be having discussions with other parties in connection with similar business relationships.

IN WITNESS WHEREOF, the parties have executed this Agreement as of the date first written above.

ABC Corporation, a _____ corporation	XYZ, a _____ corporation
By: Name: Title: President	By: Name: Title: By: Name: Title:

Sample Exclusive Patent License Agreement

between a Licensor and a Licensee
(By: Dr. Frederic Erbisch, Director [Retired],
Office of Intellectual Property, Michigan State University,
United States, e-mail: erbisch@juno.com—used with permission)

This agreement is made and entered into between _____,
a _____ established under _____
law (hereinafter called Licensor) having its principle office at
_____,
and _____ a for-profit corporation organized under the laws
of _____ (hereinafter called Licensee), having its principle
office at _____.

Witnesseth that:

1. whereas Licensor has the right to grant licenses under the licensed
patent rights (as hereinafter defined), and wishes to have the inventions covered by the licensed patent rights in the public interest;
and
2. whereas Licensee wishes to obtain a license under the licensed patent rights upon the terms & conditions hereinafter set forth:

Now, therefore, in consideration of the premises and the faithful performance of the covenants herein contained it is agreed as follows:

Article I - Definitions

For the purpose of this agreement, the following definitions shall apply:

1. Licensed Patent Rights: Shall mean:
   a. Patent   Application   Serial   No._____   filed
   _____ by _____. or
   New Plant Variety registered and protected through _____
   _____.

    b. Any and all improvements developed by Licensor, whether patentable or not, relating to the Licensed Patent Rights, which Licensor may now or may hereafter develop, own, or control.

    c. Any or all patents, which may issue on patent rights and improvements thereof, developed by Licensor and any and all divisions, continuations, continuations-in-part, reissues, and extensions of such patents.

2. Product(s): Shall mean any materials including plants and/or seeds, compositions, techniques, devices, methods, or inventions relating to or based on the Licensed Patent Rights, developed on the date of this agreement or in the future.

3. Gross Sales: Shall mean total _____ (Currency Unit) value(s) of Product(s) FOB manufactured based on the Licensed Patent Rights.

4. Confidential Proprietary Information: Shall mean with respect to any Party all scientific, business, or financial information relating to such Party, its subsidiaries or affiliates or their respective businesses, except when such information:

    a. Becomes known to the other Party prior to receipt from such first Party;

    b. Becomes publicly known through sources other than such first Party;

    c. Is lawfully received by such other Party from a party other than the first Party; or

    d. Is approved for release by written authorization from such first Party.

5. Exclusive License: Shall mean a license, including the right to sublicense, whereby Licensee's rights are sole and entire and operate to exclude all others, including Licensor and its affiliates except as otherwise expressly provided herein.

6. Know-how: Shall mean any and all technical data, information, materials, trade secrets, technology, formulas, processes, and ideas, including any improvements thereto, in any form in which the foregoing may exist, now owned or co-owned by or exclusively, semi-exclusively or nonexclusively licensed to any party prior to the

date of this Agreement or hereafter acquired by any party during the term of this agreement.

7. Intellectual Property Rights: Shall mean any and all inventions, materials, Know-how, trade secrets, technology, formulas, processes, ideas or other discoveries conceived or reduced to practices, whether patentable or not.

8. Royalty (ies): Shall mean revenues received in the form of cash and/or equity from holdings from Licensees as a result of licensing and using, selling, making, having made, sublicensing, or leasing of Licensed Patent Rights.

Article II - Grant of exclusive license

1. Licensor hereby grants to Licensee the exclusive (worldwide, option) license with the right to sublicense others, to make, have made, use, sell, and lease the Products described in the Licensed Patent Rights.

2. Licensor retains the right to continue to use Licensed Patent Rights in any way for noncommercial purposes.

3. It is understood by the Licensee that the Licensed Patent Rights were developed under _____ Grant No. _____. The _____ Government has a non-exclusive royalty free license for governmental purposes.

Article III - License Payments

1. Initial payment and royalty rate. For the licensed herein granted:
   a. Licensee agrees to pay a sign-up fee of _____ ( ).
   b. Licensee shall pay on earned royalty of _____ percent (%) of Licensee's Gross Sales of Products and fifty percent (50%) of the sublicensing receipts.
   c. Licensee shall pay an annual royalty of _____ ( ) for each leased Product.

2. Sublicenses. The granting and terms of all sublicenses is entirely at Licensee's discretion provided that all sublicenses shall be subjected to the terms and conditions of this agreement.

3. Minimum royalty: Licensee will pay Licensor, when submitting their royalty report a minimum royalty of _____ (_____) annually.

4. When a sale is made: A sale of Licensed Patent Rights shall be regarded as being made upon payment for Products made using Licensed Patent Rights.

5. Payments: All sums payable by Licensee hereunder shall be paid to Licensor in _____ (name of country) and in the currency of the _____ or in U.S. dollars.

6. Interest: In the event any royalties are not paid as specified herein, then a compound interest of eighteen percent (18%) shall be due in addition to the royalties accrued for the period of default.

Article IV - Reports, Books, and Records

1. Reports. Within thirty (30) days after the end of the calendar quarter annual period during which this agreement shall be executed and delivered within thirty (30) days after the end of each following quarter annual period, Licensee shall make a written report to Licensor setting forth the Gross Sales of Licensed Patent Rights sold, leased, or used by Licensee and total sublicensing receipts during the quarter annual period. If there are no Gross Sales or sublicensing receipts, a statement to that effect be made by Licensee to Licensor. At the time each report is made, Licensee shall pay to Licensor the royalties or other payments shown by such report to the payable hereunder.

2. Books and records. Licensee shall keep books and records in such reasonable detail as will permit the reports provided for in Paragraph 1. hereof to be determined. Licensee further agrees to permit such books and reports to be inspected and audited by a representative or representatives of Licensor to the extent necessary to verify the reports provided for in paragraph 1. hereof; provided, however,

that such representative or representatives shall indicate to Licensor only whether the reports and royalty paid are correct, if not, the reasons why not.

Article V - Marking

Licensee agrees to mark or have marked all Products made, used, or leased by it or its sublicensees under the Licensed Patent Rights, if and to the extent such markings shall be practical, with such patent markings as shall be desirable or required by applicable patent laws.

Article VI - Diligence

1. Licensee shall use its best efforts to bring Licensed Patent Rights to market through a thorough, vigorous and diligent program and to continue active, diligent marketing efforts throughout the life of this agreement.
2. Licensee shall deliver to Licensor on or before _____, a business plan for development of Licensed Patent Rights, which includes number and kind of personnel involved, time budgeted and planned for each phase of development and other items as appropriate for the development of the Licensed Patent Rights. Quarterly reports describing progress toward meeting the objectives of the business plan shall be provided.
3. Licensee shall permit an in-house inspection of Licensee facilities by Licensor on an annual basis beginning at _____.
4. Licensee failure to perform in accordance with either paragraph 1, 2, or 3 of this ARTICLE VI shall be grounds for Licensor to terminate this agreement.

Article VII - Irrevocable Judgment with Respect to Validity of Patents

If a judgment or decree shall be entered in any proceeding in which the validity or infringement of any claim of any patent under which the

License is granted hereunder shall be in issue, which judgment or decree shall become not further reviewable though the exhaustion of all permissible applications for rehearing or review by a superior tribunal, or through the expiration of the time permitted for such application (such a judgment or decree being hereinafter referred to as an irrevocable judgment), the construction placed on any such claim by such irrevocable judgment shall thereafter be followed not only as to such claim, but also as to all claims to which such instruction applies, with respect to acts occurring thereafter and if an irrevocable judgment shall hold any claim invalid, Licensee shall be relieved thereafter from including in its reports hereunder that portion of the royalties due under ARTICLE III payable only because of such claim or any broader claim to which such irrevocable judgment shall be applicable, and from the performance of any other acts required by this agreement only because of any such claims.

Article VIII - Termination or Conversion to Nonexclusive License

1. **Termination by Licensee.**
   Option of Licensee: Licensee may terminate the license granted by this agreement, provided Licensee shall not be in default hereunder, by giving Licensor ninety (90) days notice to its intention to do so. If such notice shall be given, then upon the expiration of such ninety (90) days the termination shall become effective; but such termination shall not operate to relieve Licensee from its obligation to pay royalties or to satisfy any other obligations, accrued hereunder prior to the date of such termination.

2. **Termination by Licensor.**
   Option of Licensor: Licensor may, at its option, terminate this agreement by written notice to Licensee in case of:
   a. Default in the payment of any royalties required to be paid by Licensee to Licensor hereunder.
   b. Default in the making of any reports required hereunder and such default shall continue for a period of thirty (30) days after

Licensor shall have given to Licensee a written notice of such default.

c. Default in the performance of any other material obligation contained in this agreement on the part of Licensee to be performed and such default shall continue for a period of thirty (30) days after Licensor shall have given to Licensee written notice of such default.

d. Adjudication that Licensee is bankrupt or insolvent.

e. The filling by Licensee of a petition of bankruptcy, or a petition or answer seeking reorganization, readjustment or rearrangement of its business or affairs under any law or governmental regulation relating to bankruptcy or insolvency.

f. The appointment of a receiver of the business or for all or substantially all of the property of Licensee; or the making by Licensee of assignment or an attempted assignment for the benefit of its creditors; or the institution by Licensee of any proceedings for the liquidation or winding up of its business or affairs.

3. **Effect of termination.**

Termination of this agreement shall not in any way operate to impair or destroy any of Licensee's or Licensor's right or remedies, either at law or in equity, or to relieve Licensee of any of its obligations to pay royalties or to comply with any other of the obligations hereunder, accrued prior to the effective date of termination.

4. **Effect of delay, etc.**

Failure or delay by Licensor to exercise its rights of termination hereunder by reason of any default by Licensee in carrying out any obligation imposed upon it by this agreement shall not operate to prejudice Licensor's right of termination for any other subsequent default by Licensee.

5. **Option of Licensee to convert to nonexclusive license.**

Licensee shall have the right to convert this License at the same royalty rate as for the exclusive Licensee, without right to sublicense and minimum royalties under ARTICLE III, Paragraph 3. shall not be due thereafter.

6. **Return of Licensed Patent Rights.**

Upon termination of this agreement, all of the Licensed Patent Rights shall be returned to Licensor. In the event of termination of the agreement by Licensee or said conversion of the agreement by Licensee, Licensee shall grant to Licensor a nonexclusive, royalty-free License, with right to sublicense, to manufacture, use, and sell improvements including all known-how to Licensed Patent Rights made by Licensee during the period of this agreement prior to the termination or conversion, to the extent that such improvements are dominated by or derived from the Licensed Patent Rights.

Article IX - Term

Unless previously terminated as hereinbefore provided, the term of this Agreement shall be from and after the date hereof until the expiration of the last to expire of the licensed issued patents or patents to issue under the Licensed Patent Rights under ARTICLE I. Licensee shall not be required to pay royalties due only by reason of its use, sale, licensing, lease, or sublicensing under issued patents licensed by this Agreement that have expired or been held to be invalid by an Irrevocable Judgment, where there are no other of such issued patents valid and unexpired covering the Licensee's use, sale, licensing, lease, or sublicensing; provided, however, that such non-payment of royalties shall not extend to royalty payments already made to Licensor more than six (6) months prior to Licensee's discovery of expiration or an Irrevocable Judgment.

Article X - Patent Litigation

1. Initiation. In the event that Licensor advises Licensee in writing of a substantial infringement of the patents/copyrights included in the Licensed Patent Rights, Licensee may, but is not obligated to, bring suit or suits through attorneys of Licensee's selection with respect to such infringement. In the event Licensee fails to defend any declaratory judgment action brought against any patent or patents of the Licensed Patent Rights, Licensor on written notice

to Licensee may terminate the License as to the particular patent or patents involved in such declaratory judgment action.

2. Expenses and proceeds of litigation. Where a suit or suits have been brought by Licensee, Licensee shall maintain the litigation at its own expense and shall keep any judgments and awards arising from these suits expecting that portion of the judgments attributable to royalties from the infringer shall be divided equally between Licensor and Licensee after deducting any and all expenses of such suits; provided, however, Licensor shall not be entitled to receive more under this provision than if the infringer had been licensed by Licensee.

3. Licensor's right to sue. If Licensee shall fail to commence suit on an infringement hereunder within one (1) year after the receipt of Licensor's written request to do so. Licensor in protection of its reversionary rights shall have the right to bring and prosecute such suits at its cost and expense through attorneys of its selection, in its own name, and all sums received or recovered by Licensor in or by reason of such suits shall be retained by Licensor; provided, however, no more than one lawsuit at a time shall commence in any such country.

Article XI - Patent Filings and Prosecuting

1. Licensee shall pay future costs of the prosecution of the patent applications pending as set forth in ARTICLE I, Paragraph 2. Which are reasonably necessary to obtain a patent. Furthermore, Licensee will pay for the costs of filling, prosecuting, and maintaining foreign counterpart applications to such pending patent applications, such foreign applications to be filed within ten (10) months prior to the filling date of the corresponding _____ (Country) patent application.

2. Licensor shall own improvements by the inventors. Licensee shall pay future costs of preparation, filling, prosecuting, and maintenance of patents and applications on patentable improvements

made by inventors, however, in the event that Licensee refuses to file patent applications on such patentable improvements in _____ (Country) and selected foreign countries when requested by Licensor, the rights to such patentable improvements for said countries shall be returned to Licensor.

3. Preparation and maintenance of patent applications and patents undertaken at Licensee's cost shall be performed by patent attorneys selected by Licensor; and due diligence and care shall be used in preparing, filling, prosecuting, and maintaining such applications on patentable subject matter. Both parties shall review and approve any and all patent-related documents.

4. Licensee shall have the right to, on thirty (30) days written notice to Licensor, discontinue payment of its share of the prosecution and/or maintenance costs of any of said patents and/or patent applications. Upon receipt of such written notice, Licensor shall have the right to continue such prosecution and/or maintenance on its own name at its own expense in which event the License shall be automatically terminated as to the subject matter claimed in said patents and/or applications.

5. Notwithstanding the aforegoing paragraph of this ARTICLE XI, Licensee's obligations under such paragraphs shall continue only so long as Licensee continues to have an Exclusive License under the Licensed Patent Rights and, in the event of conversion of the License to nonexclusive in accordance with ARTICLE VIII, paragraph 1. (b), after the date of such conversion:

   a. The costs of such thereafter preparation, filing, prosecuting, and maintaining of said Licensed patents and patent applications shall be the responsibility of Licensor, provided such payments are at the sole discretion of the Licensor; and

   b. Licensee shall have a nonexclusive License without right to sublicense under those of such patents and applications under which Licensee had an Exclusive License prior to the conversion.

Article XII - Notices, Assignees

1. Notices. Notices and payments required hereunder shall be deemed properly given if duly sent by first-class mail and addressed to the parties at the addresses set forth above. The parties hereto will keep each other advised of address changes.

2. Assignees, etc. This Agreement shall be binding upon and shall inure to the benefit of the assigns of Licensor and upon and to the benefit of the successors of the entire business of Licensor, but neither this agreement nor any of the benefits thereof nor any rights thereunder shall, directly or indirectly, without the prior written consent of Licensor, be assigned, divided, or shared by the Licensor to or with any other party or parties (except a successor of the entire business of the Licensor).

Article XIII - Miscellaneous

1. This agreement is executed and delivered in _____ (Country) and shall be constructed in accordance with the laws of the Government of _____.

2. No other understanding. This agreement sets forth the entire agreement and understanding between the parties as to the subject matter thereof and merges all prior discussions between them.

3. No representations or warranties regarding patents of third parties. No representations or warranty is made by Licensor that the Licensed Patent Rights manufactured, used, sold, or leased under the Exclusive License granted herein is or will be free of claims of infringement of patent rights of any other person or persons. The Licensor warrants that it has title to the Licensed Patent Rights from the inventors.

4. Indemnity. Licensee shall indemnify, hold harmless, and defend Licensor and its trustees, officers, employees, and agents against any and all allegations and actions for death, illness, personal injury, property damage, and improper business practices arising out of the use of the Licensed Patent Rights.

5. Insurance. During the term of this agreement, Licensee shall maintain the following insurance coverage:
   a. Commercial general liability with a limit of no less than one million dollars ($1,000,000.00, option) each occurrence. Such insurance shall be written on a standard ISO occurrence form or substitute form providing equivalent coverage.
   b. Professional liability of no less than one million dollars ($1,000,000.00, option) each occurrence.
   c. Workers' compensation consistent with statutory requirements. Certificates of insurance shall be provided to Licensor upon request and shall include the provision for 30-day notification to the certificate holder of any cancellation or material alteration in the coverage.
6. Advertising. Licensee agrees that Licensee may not use in any way the name of Licensor or any logotypes or symbols associated with Licensor or the names of any researchers without the express written permission of Licensor.
7. Confidentiality. The parties agree to maintain discussions and proprietary information revealed pursuant to this agreement in confidence, to disclose them only to persons within their respective organizations having a need to know, and to furnish assurances to the other party that such persons understand this duty on confidentiality.
8. Disclaimer of Warranty. Licensed Patent Rights is experimental in nature and it is provided **without warranty or representations of any sort, express or implied, including without limitation warranties of merchantability and fitness for a particular purpose of non-infringement**. Licensor makes no representations and provides no warranty that the use of the Licensed Patent Rights will not infringe any patent or proprietary rights of third parties.

In witness whereof, the parties hereto have caused this agreement to be executed by their duly authorized representatives.

The effective date of this agreement is _____, 20_____.

Licensor _____

Name: _____

Title: _____

Licensee: _____

Name: _____

Title: _____

SAMPLE OPTION TO LICENSE AGREEMENT

(By: Dr. Frederic Erbisch, Director [Retired], Office of Intellectual
Property, Michigan State University, United States,
e-mail: erbisch@juno.com—used with permission)

This Agreement is made and entered into between the _____ ,
a research establishment under _____ laws (hereinafter
"Grantor") having its principle office at, _____ and
_____ a company
organized under the laws of _____ (hereinafter "Grantee"),
having its principle office at _____.

1. Grant of Option. In consideration of payment of the Option
   Price by the Grantee to the Grantor, receipt of which the Grantor
   acknowledges, the Grantor grants the Grantee an exclusive option
   to obtain a license from the Grantor to the Optioned Rights, in
   accordance with this Option Agreement.

2. Definitions.
   a. Option Price means that amount which the Grantor and the
      Grantee agree shall be paid for the Optioned Rights within the
      Term. The Option Price shall be _____
      (_____).
   b. Optioned Rights means the intellectual property herein described
      as: _____

  c. Term means that period of time, which Grantor and Grantee agree shall allow Grantee to evaluate the Optioned Rights. The Term shall be _____ to _____.

3. Exercise of the Option. The Grantee may exercise its option at any time prior to expiration of the Term by giving written notice signed by the Grantee to the Grantor at its address stated above. The notice must be personally delivered or postmarked before the expiration of the Term.

4. Confidentiality. The parties agree to maintain discussions and proprietary information revealed pursuant to the Option Agreement in confidence, to disclose them only to persons within their respective companies having a need to know, and to furnish assurances to the other party that such persons understand this duty of confidentiality.

5. Conditions to License. In the event the Grantee elects to exercise its option, execution of a license agreement, fulfillment of the following conditions to license shall occur within thirty (30) days after the Grantor receives the notice that the Grantee is exercising the option. The Grantee shall pay to the Grantor the following amount and meet the following requirements at the time it enters into a license agreement with the Grantor:

  a. Pay an initial license fee of _____ ( ) and a royalty rate not to exceed _____ percent (%) of the net sales of Products (as defined in the License Agreement), and

  b. Provide the Grantor with a preliminary business plan acceptable to the Grantor that describes the steps proposed by the Grantee to commercialize the Optioned Rights.

6. Terms of License. Terms and conditions of the license agreement will be negotiated in good faith so as to result in a license acceptable to both parties substantially in the form of Exhibit A.

7. Failure to Exercise Option or to Close. If the Grantee fails to exercise its option properly before expiration of the Term or fails to meet the conditions to license and enter into a license within the time allowed, this Option Agreement shall terminate and the Grantor

may retain the Option Price and shall have no further obligation to the Grantee.

8. Assignment. This Option Agreement shall bind and benefit the parties' successors and assigns. Neither party may assign rights under this Option Agreement without the prior written consent of the other party.

9. Entire Agreement; Amendment. This Option Agreement contains the entire agreement of the parties with respect to the transaction described in this Option Agreement, and no prior or simultaneous oral or other written representations or promises shall be a part of this Agreement or otherwise effective. This Option Agreement may not be amended or released, in whole or in part, except by a document signed by both parties.

10. Indemnity. Grantee shall indemnify, hold harmless, and defend Grantor and its trustees, officers, employees, and agents against any and all allegations and actions for death, illness, personal injury, property damage, and improper business practices arising of the Optioned Rights.

11. Warranty. Optioned Rights is experimental in nature and it is provided without warranty or representations of any sort, express or implied, including without limitation warranties of merchantability and fitness for a particular purpose of noninfringement. Grantor makes no representations and provides no warranty that the use of the optioned Rights will not infringe any patent or proprietary rights of third parties.

12. Interpretation. The paragraph headings used in this Option Agreement are provided for convenience of reference only and shall not be used to interpret the provisions of this Option Agreement. In the event any provision of this Option Agreement proves to be illegal or unenforceable, the remaining provisions of this Option Agreement shall be interpreted as if such illegal or unenforceable provision were not a part of this Option Agreement.

13. Law. This Option Agreement is executed and delivered in the _____ and shall be constructed in accordance with the laws of the Government of _____.

To evidence their agreement to the foregoing terms and conditions, the Grantor and the Grantee have executed this Option Agreement below.

Grantor: _____   Grantee: _____

By: _____   By: _____

Signature: _____   Signature: _____

Title: _____   Title: _____

Date: _____   Date: _____

SAMPLE CONSULTING AGREEMENT WITH LICENSEE

By Dr. Robert Yonover

This Agreement is made and entered into as of this _____ day of _____, by and between SEE/RESCUE® Corporation, a technology promotion, government relations, consulting, and international business development firm having its office at (enter address) and Xyz Corp., having its principal office at (enter address).

Witnesseth:

WHEREAS, SEE/RESCUE® Corporation (hereinafter "SRC") wishes to provide government relations, consulting, and intellectual input to Xyz Corp.

WHEREAS, Xyz Corp. wishes to obtain the services of SRC as set forth in accordance with this agreement.

NOW THEREFORE, in consideration of the foregoing recitals and the mutual covenants, terms, conditions, and agreements hereafter provided, the parties mutually agree as follows:

1. ***Effective Date.*** This Agreement shall take effect on _____ and shall continue through _____. Following the expiration of the initial term hereof, this Agreement shall

automatically be renewed for successive one (1) year terms, unless either party gives the other written notice of intent not to renew at least thirty (30) days prior to the expiration of the then-existing term.

2. ***Compensation.*** SRC shall provide consulting and intellectual input to Xyz Corp. As compensation for its consulting and intellectual input, Xyz Corp. agrees to pay $xx annually, payable in equal monthly installments of $xx per month for the services of SRC. Such payment shall be due on the fifteenth day of each month beginning on September 15, 2007.

3. ***Indemnification.*** Xyz Corp. shall indemnify and hold SRC harmless from and against any and all liability, loss, damage, cost, or expense (including reasonable attorney's fees) resulting from the acts or omissions, negligence or intentional wrongdoing of Xyz Corp. SRC shall indemnify and hold Xyz Corp. harmless from and against any and all liability, loss, damage, cost, or expense resulting from the acts or omissions, negligence or intentional wrongdoing of SRC.

4. ***Independent Contractor.*** SRC will act as an independent contractor in the performance of its consulting and intellectual input under this Agreement. SRC is not responsible for the acts of Xyz Corp. or representations made by Xyz Corp. upon which SRC acts in providing consulting and intellectual input under this Agreement.

5. ***Assignment.*** This agreement may not be assigned by Xyz Corp. or SRC without the prior written consent of both parties.

6. ***Nondisclosure.*** SRC agrees to hold all Xyz Corp. proprietary information and intellectual property in trust and confidence. SRC agrees not to publish, disseminate, or disclose such information without the prior written consent of Xyz Corp.

7. ***Applicable Law.*** This Agreement shall be construed, interpreted, and governed by and in accordance with the laws of (state) without regard to the principles of conflicts of laws.

8. ***Notices.*** Notices shall be sent to the parties at the addresses first set forth above. Any person to whom notice may be given hereunder may from time to time change said address by written notice

through the U.S. mail service or equivalent service such as Federal Express.

9. ***Severability.*** If a court of competent jurisdiction declares that any term or provision of this agreement is invalid or unenforceable then: 1) the remaining terms and provisions shall be unimpaired, and 2) the invalid or unenforceable term or provision shall be deemed replaced by a term or provision that is valid and enforceable and that comes closest to expressing the intention of the invalid or unenforceable terms or provisions.

10. ***Entire Agreement.*** This Agreement constitutes the entire agreement among the parties with respect to the matters contained herein. Any modification or amendment to this Agreement must be made only by written mutual consent of both parties.

IN WITNESS WHEREOF, the parties have executed this Agreement, the day, month, and year first above written.

Xyz Corp.
By: ⎯⎯⎯⎯⎯⎯⎯⎯⎯⎯⎯⎯⎯⎯⎯⎯⎯⎯⎯

SEE/RESCUE® Corporation (SRC)
By: ⎯⎯⎯⎯⎯⎯⎯⎯⎯⎯⎯⎯⎯⎯⎯⎯⎯⎯⎯
Dr. Robert Yonover

SAMPLE MARKETING LETTER

(from the National Inventor Fraud Center, www.inventorfraud.com)

[Your Name]
[Your Address]
[Your Phone #]

[Date]

[Contact Person's Name]

[Manufacturer's Name]
[Manufacturer's Address]
Re: [NAME OF YOUR INVENTION]
Dear [Contact Name of Manufacturer]:
I am the inventor of an invention titled "[INVENTION TITLE]." I filed a utility patent application for my invention on [filing date] through the law firm of [name of law firm].

I selected your company because you manufacture [identify their products that are similar to your invention]. I believe the [intention title] will benefit your company in many different ways. [Identify how your invention fits within the manufacturer's product line].

Briefly stated, my invention comprises [briefly identify major components]. The unique features of my invention are [unique features]. The advantages of the [invention title] are [state advantages].

I have enclosed a brochure of my invention which briefly displays the key components for your review. You may also visit my Web site at (your Web site) for more detailed information about the [invention title].

I am willing to consider selling or licensing the patent rights to my invention. If you are potentially interested in the [invention title], please contact me anytime at [your phone number].

I look forward to your response.

Very truly yours,
(Your name)

Enclosure: Brochure

(DISCLAIMER: While the authors and publisher have made every effort to make this book as complete and accurate as possible, they make no representations or warranties with respect to accuracy or completeness of the contents and specifically disclaim any implied warranties. The advice and strategies contained herein may not be suitable for your situation. The author and the publisher are not engaged in rendering legal, accounting, or other professional services and you should consult with a professional where appropriate. Neither the authors nor the publisher shall be liable for any loss of profit or any other commercial damages, including but not limited to special, incidental, consequential, or other damages.)

ACKNOWLEDGMENTS

MANY PEOPLE CONTRIBUTED THEIR KNOWLEDGE AND experience to enrich this book. Thank you to Bill Wolfsthal, Kathleen Go, Heidi Bollich, and the editors at Skyhorse Publishing, and to Neil Salkind of the Salkind Agency and Studio B for believing in the project. We are honored and grateful to have a foreword from Louis Zamperini. Thank you to Kyle Kolker for a great book cover and to Micah Fry for such cool illustrations. Many, many thanks to Will Crowe for long hours of research, good ideas, fact-checking, proofreading, and indexing. Thank you to the experts who graciously agreed to read the book and provide endorsements.

We thank all the inventors who shared their inspirational stories with us, and the experts who allowed us to quote their very useful advice, including: Nate Ball, Owen Baser, Paul Berman, Ron Bessems, Brian Boothe, Amanda Budde-Sung, Susanne Chess, Peter Cooper, Craig Coppola, Don Debelak, Leonard Duffy, Dr. Frederic Erbisch, Bob Evans, Shawn Frayne, Sue and Kathryn Gregory, Cameron Gunn, Professor Hugh Herr, Donna, Kristin, and Frank Hrabar, John Jantsch, David Jisa, Christopher Johnson (Name Inspector), Annette Kalbhenn, Guy Kawasaki, Peter Kay, Stephen Key, Tim Kehoe, Vadim Kotelnikov, Andrew Krauss, Jim Lowrance, Michael Michalko, Bob Merrick, Michael Neustrel, OceanLink, Mark Ott, David Parrish, Joe Pedott, Barbara Pitts, David Pressman, Boris Rubinsky, Bryan Schmidt, Dr. Charles Steilen, Rich Stim, Mike Sykes, SUNGRI,

Barbara Uebelacker, ThinkGeek, Ken Thompson, Ray Vance, Kelydra Welcker, Sue Wyshynski, and Ed Zimmer.

Although lawyers often get a bad rap, I wanted to send out a special thanks to a few that have had my back from the beginning, through thick and thin: Christine Weger and Richard Diehl, Diehl & Weger, Honolulu, for general/accounting (cweger@hawaii.rr.com); Josefino P. De Leon, Shlesinger, Arkwright & Garvey, Alexandria, VA, for patents/IP (jpdeleon@sagllp.com); and Sandra Slon, Troy & Gould, Los Angeles, for corporate/license agreements (sslon@troygould.com).

On the professional front, I want to thank my science mentors, Dr. Mike Sommer and Dr. John Sinton, for teaching me the scientific method and providing me with the opportunity to flirt with volcanoes in Hawaii and the rest of the world. Bertil Werjefelt taught me the ropes on inventing and Nora Feuerstein showed me how it's done in the big leagues. Thanks to Kendall Kikuyama and Rescue Technologies Corporation for manufacturing and distributing RescueStreamer, and David Rensin, who showed me the light on how to invent a book and provided me with ongoing momentum.

On a personal front, Carl Middelmann, Wayne Giancaterino, Jack Bredin, Mike Armstrong ("Army"), Jordon Cooper, and Chuck Mitsui were always willing to listen to my rants and make sure I got my required dose of North Shore surf. Dick Brewer, Jack Reeves, Tommy Nellis, and Jim Porteus kept me quivered up. Pat Caldwell kept me in the know on the surf forecast and Ken Braeseke, Mike Carbone, and Perry Farrell were always there to defend the Haulover turf. Steve and Gail Carr made it possible for me to stay afloat, even when the boat flipped over!

Neal Yonover, my big brother, has been a partner in crime since the real beginning, critiquing and providing marketing hints for the earliest of inventions, even my electric nose picker (in part to avoid parental havoc)!

I also thank my father, my right-hand man and the voice of "every man," who continues to provide me with the sounding board for early-stage ideas, concepts, and fantasies. My mother, for infusing a never-ending creativity bug in us kids, and my sister, Ann, for providing ongoing perspective from a real artist.

Finally, a massive thank-you to my soul mate Cindy, who has been my shipmate in every storm we have collectively encountered and was an integral part of the team that created our children, Jesse and Kera—the ones who always show me the fresh, unadulterated perspective on ideas and life in general!

SELECTED SOURCES

About.com (www.investors.about.com/library/weekly/aa091297.htm)

BNET Today (findarticles.com/p/articles/mi_m0DTI/is_10_28/ai_66239772)

Ecosistemaurbano.org (ecosistemaurbano.org/?p=1933#more-1933)

The Engines of Our Ingenuity (www.uh.edu/engines/engines.htm, www.uh.edu/engines/epi1120.htm)

Florida Trend magazine (www.floridatrend.com/article.asp?page=5&aID=47796, www.floridatrend.com/article.asp?aid=47843)

Good Clean Tech.com (www.goodcleantech.com/2007/10/inventor_ shawn_frayne_creates.php)

GuyKawasaki.com (www.guykawasaki.com/about/index.shtml)

The Heinz Awards (www.heinzawards.net/recipients/hugh-herr)

Kids News Room (www.kidsnewsroom.org/newsissues/051807/index .asp?page=Weekly2)

Neustel Law Offices (www.patent-ideas.com, www.patent-ideas.com/patents.htm)

NewScientist.com (www.newscientist.com/article/mg19826545.500-cellphone-scanner-could-screen-for-cancer.html)

Popular Mechanics.com (www.popularmechanics.com/science/earth/4212739.html, www.popularmechanics.com/technology/industry/4225816.html?page=8&series=37)

Shop4patents.com (www.shop4patents.com/articles/tag/spherical-safety-seat)

Smithsonian.com (www.smithsonianmag.com/arts-culture/object-chiapet-200712.html)

Sunrgi.com (www.sunrgi.com/press20080429.html)

T2design.com (www.t2design.com/entrepreneur.html)

Time.com (www.time.com/time/specials/packages/article/0,28804,1852747_1854195_1854170,00.html)

Welcome to the National Zoo|FONZ website (www.Nationalzoo.si.edu/publications/zoogoer/1999/4/designsfromlife.cfm)

Weston Solutions.com (www.westonsolutions.com/about/news_pubs/press_releases/greengrid.htm)

WikiAnswers.com http://wiki.answers.com

Wristies.com (www.wristies.com/KK_Wristies_Inventor_s/152.htm)

INDEX

Advance Praise for Hardcore Inventing

"*Every page provides sound practival advice and inspiring real-life examples.* Hardcore Inventing*'s can-do approach and action-based ideas will make inventors reach for their got-to-do list.*"
  —*Guy Kawasaki, one of the original Apple/MacIntosh evangelists, author of* Reality Check *and co-founder of Alltop.com*

"*This book is definitely required reading for anyone interested in inventing for profit. Rob's book will arm you with special knowledge based on his original 'IP³' approach, which in my opinion will keep you on track to becoming a successful inventor. I heartily recommend Rob's book to all beginning and struggling inventors.*"
  —*Robert G. Merrick, successful inventor/entrepreneur and author of* Stand Alone, Inventor!

"*Just like the RescueStreamer® and his many other creations,* Hardcore Inventing *provides exacting methodology, relentless enthusiasm, and all the sage guidance that anyone who wants to birth a great idea will find in this excellent and inspirational book.*"
  —*David Rensin, author of* All For A Few Perfect Waves: The Audacious Life and Legend of Rebel Surfer Miki Dora

"Hardcore Inventing *has it all! An easy to use, informative and inspirational guide on how to take your idea and turn it into success. Using* Hardcore Inventing, *our seemingly crazy idea went all the way! Dreams can come true... with some help from Dr. Rob Yonover and his valuable book.*"
—*Dr. Matthew Boyd, BSc, BVMS, director of Aroomafresh LLC and winner of the 2008 Hawaii Pacific University Venture Challenge*

"*As an entrepreneur and creative thinker Dr. Rob Yonover's inventive thinking and guerilla marketing is nothing but amazing. He inspires and forces each of us to think out of the box with endless passion and smart inventive strategies.*"
—*Patrick Sean Flaherty, founder of Brandhalo.com*

"*Rob has a great non-nonsense communication style that appeals to people [who] want to cut through the fat and get to the bone. If you want to create a life like Rob's, that's based on inventing great products, is personally satisfying, and stays far away from cubicle hell, this book is for you.*"
—*Peter Kay, Serial Tech Entrepreneur, YourComputerMinute.com*

"*This book is a concise and brilliantly simple approach, perfectly organized with salient examples, sprinkled with wit that makes it impossible to set down. Any aspiring inventor or rational dreamer would be foolish not to embrace* Hardcore Inventing *as the primary guide for success. It is a logical check list to success based on great ideas and determination to succeed.*"
—*Ralph J.W.K. Hiatt, Col. (Ret.), Army Special Operations*